Exmoor ponies
and the livestock genesis of the horse

Exmoor ponies
and the livestock genesis of the horse

Helmut Hemmer

© 2024 Prof.Dr. Helmut Hemmer
Verlag: BoD • Books on Demand GmbH, In de Tarpen 42, 22848 Norderstedt
Druck: Libri Plureos GmbH, Friedensallee 273, 22763 Hamburg

ISBN: 978-3-7597-8753-8

Contents

Foreword 1

Foreword to the English edition 3

1 The Exmoor pony 4

1.1 External appearance 4
1.2 Character 8
1.3 Habitat 8
1.4 History 9

2 The controversy 12

3 Domestication – livestock genesis 17

3.1 Definition of terms 17
3.2 Preadaptations 24
3.3 Systematics, taxonomy and nomenclature 31
3.4 The decision question 36

4 Wild horses 38

4.1 Knowledge from palaeontology 38
4.2 Knowledge from archaeozoology 41
4.3 Knowledge from molecular genetics 42
4.4 The tarpan 43
4.5 The Przewalski's horse 45

Contents

5   Domestic horses   50

    5.1 History   50
    5.2 Diversity   52
    5.3 Pseudo-wild horses   54

6   Behavioural diagnostics   61

    6.1 Introduction   61
    6.2 Individual distance   62
    6.3 Coordination   65
    6.4 Attentiveness   67
    6.5 Movement intensity   68

7   The jigsaw puzzle   70

    7.1 Molecular genetic evidence   70
    7.2 Brain size   71
    7.3 Mane   72
    7.4 Diagnostic structures   72
    7.5 European mountain pony – *Equus ferus arcelini*   79

8   The livestock genesis of the horse   83

9   Outlook   89

10   Register   92

To my adopted daughter Angela Carrasco

Foreword

Anyone who deals with domestic animals and their origins must also deal with the horse. Anyone who studies the horse comes up against the question of the Exmoor pony. Is this small horse which still lives more or less wild in its original homeland, the last survivor from the once huge herds of European wild horses? Or is it one of the many breeds of western and northern European ponies, that were differentiated from the first European domestic horses by breeding from the Bronze Age onwards, but above all in the last two millennia?

The author of this book initially intended to address this problem more casually in the course of clarifying a comparative issue for cattle, which was being discussed at the time, particularly in England. The aim was to investigate the Chillingham Park cattle, which had repeatedly been compared to small aurochs, albeit with different colour. In the summer of 1985, he travelled to England to inspect the original herd living freely in the extensive grounds of Chillingham castle in Northumberland and to make behavioural observations, as well as to study skull material in the Hancock Museum in Newcastle and the British Museum of Natural History in London.

After the cattle question had been resolved fairly quickly in favour of a clear domestic animal finding, discovering more about the Exmoor pony proved impossible. At the time, the British Natural History Museum did not have a single skull of this small horse in its extensive collections. Whipsnade Park Zoo, which harboured several British pony breeds side by side had everything but Exmoor ponies. The author previously only had contact with this small horse in Michael Schäfer's extensive paddocks in Erdinger Moos, although at that time his main interest was in the conservation breeding of the Portuguese Sorraia pony there.

After the failure in England, he then made the Exmoor pony a topic of discussion in his Mainz team. In the course of eight years, the authors Bernadette Riediger (1995), Manuela Jendrosch (1997) and Karin Siebert (2003) wrote three consecutive and complementary diploma theses on behavioural comparisons with other horse forms. These studies remained unpublished as such, but their contents are now evaluated as a decisive basis for clarifying the problem of categoryisation in this book.

At this point the author is grateful to Gisa Heinemann from the Biodiversity Museum of the University of Göttingen for numerous photos and scans of the upper dentition and metacarpal measurements from the large Exmoor pony collection there, which can justifiably be regarded as a museum reference collection for this horse. The author also likes to thank Bärbel Fiedler from the Senckenberg Research Station for Quaternary Palaeontology in Weimar for her support in obtaining literature. For the numerous excellent photographs of Exmoor ponies in their original habitat on Exmoor, which were made available for this book, special thanks are due to Wolfgang Frey (Germering), who documented his great interest in the wild animal – domestic animal nature of this horse at an early stage by collecting extensive photographic material. All photos not labeled with a name were taken by the author himself.

Foreword to the English edition

Following the publication of the original German edition of this book, Sue Baker, leading expert on the Exmoor pony, argued that it should be published in English, so that it could be read in the animal's home country. She wrote "I think the future for the Exmoor pony will depend upon full acceptance of them being a race rather than a breed, generically wild game not man-made livestock". This acceptance can only be achieved if it spreads to where the Exmoor pony is at home and from where its future is controlled worldwide through the Exmoor Pony Society. Without doubt, this book must be able to be studied on Exmoor itself. Only an English version can provide the basis for this.

The author was happy to take up this suggestion. In order to keep the production costs at a reasonable level in view of the expected limited readership, he used translation software as the basis for the English text. As a result of the original version, which is quite condensed in many cases, and above all because of the centrally used new German word *Viehwerdung*, this automatic translation produced absolutely incomprehensible results in many passages. In all such places, the author revised the text himself and then submitted it to Sue Baker in order to eliminate any linguistic problems that had arisen. Special thanks go to her for this crucial help. To make print and sale possible for the English-speaking countries by the original German publishing house, a certain diminution of the print quality could not be avoided, before all a shift from colour to black-and-white illustrations.

1 The Exmoor pony

1.1 External appearance

Exmoor ponies are small horses with an adult shoulder height of between 112 and 135 cm (11.0-13.2 hands), on average 124.5 cm (12.2 hands). The broad and deep rump is supported by relatively short, strong legs. The basic stockiness is subject to a certain variability from rather slender to rather stocky. There are also differences in head shape between individual herds.

Figure 1.1: The colour of the Exmoor pony varies from light brown to dark brown with basically the same colour pattern. Characteristic features are the mealy muzzle and a bulging eye ring with light coloured hair, which gives the impression that the eyes are protruding ("toad eyes"). This and all other pictures of Exmoor ponies in this book were taken in Exmoor National Park. (Photo Wolfgang Frey)

Figure 1.2: Lighter colour variant of the Exmoor pony. The brand of the thigh demonstrates the procedure for the Exmoor pony corresponding to a studbook-controlled domestic horse breed. (Photo Wolfgang Frey)

While the summer coat has a certain metallic sheen, the winter coat has a lustreless top coat over a very dense, thick, greasy undercoat and allows rain and snow water to run off. The mane hangs to one side of the neck and thus also helps the rain to run off. The forelock can become so long and thick that the small ears almost seem to disappear.

The colour of the coat ranges from light brown to a very dark brown which is more or less lightened on the under chest and belly, especially in the rear area, sometimes as high as the flanks, as well as on the hind cheeks. The mealy muzzle and the equally light-coloured eye rims stand out very conspicuously against the brown. This makes the eyes appear prominent. On the legs, the colour of the metapodia turns blackish. Apart from the wide brown range, the coat colour of the Exmoor ponies is of high uniformity. In earlier times there were also grey ponies and possibly some other colours in the population.[1]

Figure 1.3: Deep dark variant of the Exmoor pony. A comparison with the previous picture also demonstrates the variability of the body structure. (Photo Wolfgang Frey)

---

[1] Essential literature on the appearance of the Exmoor pony, basis of this chapter:
Baker, S. (2008): Exmoor-Ponies – Survival of the Fittest – A Natural History. Revised and updated edition. Chippenham (Exmoor Books).
Baker, S., Creig, C., Macgregor, H., & Swan, A. (1998): Exmoor Ponies – Britain's prehistoric wild horses? Brit. Wildlife, 9: 304-313.
Dent, A. A. (1970): The pure bred Exmoor Pony. Williton (Cox).
Nissen, J. (1998): Enzyklopädie der Pferderassen: Europa, Vol, 2. Stuttgart (Franckh-Kosmos).
Schäfer, M. (2000): Handbuch Pferdebeurteilung. Stuttgart (Franckh-Kosmos).
Willman, R. (1999): Das Exmoor-Pferd: eines der ursprünglichsten halbwilden Pferde der Welt. Natur und Museum, 129, 12: 389-407.

This colour distribution, with its typical lightening and darkening, as well as the mealy muzzle, corresponds fundamentally to that of the Dzungarian wild horse or Przewalski's horse, with the exception of the absence of a clear eel line on the middle of the back, although the entire colour variation range of the latter is shifted to a considerably lighter shade. This type of colouring, with or without the mealy muzzle, which exists in both brown and grey variants ("mouse-gray"), can be regarded as the original wild colour of the horse. It has been preserved in a number of domestic horse breeds that were never subjected to intensive breeding and is documented in Western European cave paintings many thousands of years before domestication.

Figure 1.4: Newborn Exmoor ponies initially have a lighter coat, whereby the mealy muzzle and the white eye ring already stand out in the typical manner. The legs are also whitish in colour at this very young age. Particularly on the flanks, a very delicate zebra striping may appear. (Photo Wolfgang Frey)

1.2 Character

Exmoor ponies living in the wild are characterised by great shyness and very prudent behaviour with regard to current requirements of their environment. Unused to humans, adults caught in the wild are almost impossible to control until sensitively tamed. Their robustness, hardiness, resilience and endurance are remarkable, as is their longevity, which is unusual for horses.[2] Special character traits are their energetic temperament coupled with a friendly, good-natured character and high intelligence. On Exmoor they say "If you can ride an Exmoor, you can ride everything". Although these ponies are certainly suitable as children's ponies, they are less suitable for children who are new to riding, as they are generally too strong, too energetic and too intelligent.[3]

1.3 Habitat

The Exmoor pony originates from the moorland of the Exmoor area above the cliffs to the south of the Bristol Channel in south-west England. Its native habitat here is partly rocky, barren, rough, hilly moorland with only sparse tree-cover, largely characterised by grassland, heather, gorse and hawthorn. They are also ferns, marshy areas and bogs. Deep valleys are wooded. In addition to the ponies, there is a large population of red deer. The area, which is divided into common land and farmland, is also grazed by sheep and cattle. Winters in this landscape are cold and very rainy.[4] The ecological plasticity of the Exmoor pony is extraordinarily high. Where woodland and open land are available at the same time, it shows a preference for the latter, but does not avoid the woodland in any way and stays in it for a long time.[5]

---

[2] Many anecdotal examples that shed light on their special nature for horses can be found in: Baker, S. (2017): Exmoor Pony Chronicles. Wellington (Halsgrove).

[3] Nissen, J. (1998): Enzyklopädie der Pferderassen: Europa, Vol, 2. Stuttgart (Franckh-Kosmos).

[4] For descriptions of the habitat, see the literature listed in footnote [1].

[5] Rödde, S. M.-C. (2015): Verhalten und Raumnutzung von Exmoorponys im Reiherbachtal (Solling). PhD Thesis, Georg August-University Göttingen.

Figures 1.5-1.6: Exmoor ponies in their habitat in Exmoor National Park. (Photos Wolfgang Frey)

1.4 History

The first documentary mention was in 1086, when William the Conquerer ordered a list to be drawn up of the property and livestock in Exmoor. This list documents a large number of horses. From William

II (1056-1100) at the latest, Exmoor was a royal hunting ground, which severely restricted the rights of the local inhabitants to graze their own livestock there. A campaign by Henry VIII to allow only large stallions for breeding throughout the country came to nothing in this remote area. As late as the 18th century, the Exmoor ponies were treated the same as other game in terms of naming. At the beginning of the 19th century, the area, which until then had belonged to the Crown, was privatised. The former royal warden acquired part of it and set up a stud farm with selected animals on his property for the purpose of long-term conservation of the original Exmoor pony in its ancestral landscape. Other landowners gradually joined in this endeavour.

Based on this core herd and the interest of farmers and their heirs at the time, the continuous existence of the purebred Exmoor pony was prevented from being jeopardised by the extensive experimentation with crossbreeding of all kinds to "improve" the animals in the 19th century. At the beginning of the 20th century, the Exmoor Pony Society was founded, modeled on a regional sheep breeders' association.

A considerable reduction in genetic heritage was a direct consequence of the Second World War. The majority of animals living on Exmoor were captured and stolen for the sale of meat to the hungry urban population. Of the core herd only a dozen ponies escaped the massacre, and together with other herds only around 50 animals in total survived. In the decades following the war, ponies were repeatedly exported, eventually establishing several independent herds outside of Great Britain. Besides the central aim of conservation, Exmoor ponies have since been used in several countries, including Germany, for landscape management grazing projects.[6]

---

[6] The history of the Exmoor pony:
Baker, S. (2008): Exmoor-Ponies – Survival of the Fittest – A Natural History. Revised and updated edition. Chippenham (Exmoor Books).
Baker, S. (2017): Exmoor Pony Chronicles. Wellington (Halsgrove).
Baker, S., Creig, C.,Macgregor, H., & Swan, A. (1998): Exmoor Ponies – Britain's prehistoric wild horses? Brit. Wildlife, 9: 304-313.
Nissen, J. (1998): Enzyklopädie der Pferderassen: Europa, Vol, 2. Stuttgart (Franckh-Kosmos).

Figure 1.7: Exmoor pony in its native habitat in Exmoor National Park above the Bristol Channel. (Photo Wolfgang Frey)

2 The controversy

The opinions of the inhabitants of the Exmoor in earlier centuries regarding the identity of the small horses living there have not been definitely recorded. Individual names speak more in favour of them being regarded as game rather than livestock, especially as it was a hunting ground of the Crown. From the 19th century at the latest, they were regarded as one of other British and northern European pony breeds with very special characteristics that were worth preserving, but which individual farmers would also have liked to "improve" through breeding. However, all attempts to do so proved to be unsuccessful.[7]

Figures 2.1-2.4: The first alternative: Exmoor ponies belong with British and Nordic small horse breeds. Exmoor pony in Exmoor National Park (top left; photo Wolfgang Frey), New Forest pony (top right), Dartmoor pony (bottom left), both in Whipsnade Park Zoo, Icelandic horse in Iceland (bottom right).

---

[7] Baker, S. (2017): Exmoor Pony Chronicles. Wellington (Halsgrove).
Nissen, J. (1998): Enzyklopädie der Pferderassen: Europa, Vol, 2. Stuttgart (Franckh-Kosmos).

In the middle of the 19th century, the scientific name *Equus caballus britannicus* Sanson, 1869, appeared, which was later mistakenly attributed to the Exmoor pony. However, it referred to various horses of the English and French coastal regions, without mentioning the animals from Exmoor.[8] In any case, as a name of a breed of domestic animal, it could not be valid in zoological nomenclature (Chapter 3.3).

Towards the end of the first half of the 20th century, the controversial view was first expressed that the Exmoor pony might not actually be a domestic horse at all, but a surviving wild horse.[9] Subsequently, osteological studies came into play, which initially compared current British ponies with Pleistocene fossil finds. On this basis, the Speed/Ebhardt pony and horse types were formulated, with the Exmoor pony set as type I. It represents the anatomy of British wild horses long before the time of domestic animal emergence. All four horse types (hippological horse types) are to be understood in terms of functional morphology and are based on different combinations of structural features and anatomical characteristics of the lower jaw and metapodial bones that stand out in X-rays.[10] In the second half of the 20th century, they played a major role in assessing the breeding background of horses,[11] but are of rather limited value in clarifying evolutionary relationships.[12]

---

[8] Sanson, M. A. (1869): Nouvelle détermination des espèces chevalines du genre *Equus*. Comptes rendus hebdomadaires des séances de l'Académie des sciences, 69: 1204-1207.
Heptner, V. G., Nasimovich, A. A., & Bannikov, A. G. (1988): Mammals of the Soviet Union, Vol. 1, Artiodactyla and Perissodactyla. Washington (Smithsonian).

[9] Best, M. G. S. (1947) in: Best, M.G.S. & Etherington, M. G. (Eds.): The little horses of Exmoor. Cited after: Baker, S. (2017): Exmoor Pony Chronicles. Wellington (Halsgrove).

[10] Speed, J. G. (1951): Horses and their teeth. The Journal of the Royal Army Veterinary Corps, 22, 4. Reprinted in Speed, J. G. & Speed, M. G. (Eds./1977): The Exmoor-Pony, 63-69. Chippenham (Countrywide Livestock Ltd).
Ebhardt, H. (1962): Ponies und Pferde im Röntgenbild nebst einigen stammesgeschichtlichen Bemerkungen dazu. Säugetierkundliche Mitteilungen,10: 145-168.
Ebhardt, H. (1964): Zusammenhänge zwischen Zahnbau, Zahnstellung und Kieferbau bei Pferdeunterkiefern. Säugetierkundliche Mitteilungen, 12: 145-155.
Ebhardt, H. (1967): Das Überleben des Ponytyps I bis in die Nacheiszeit und sein Auftauchen in der frühen Domestikation. Säugetierkundliche Mitteilungen, 15: 18-35.

[11] Schäfer, M. (2000): Handbuch Pferdebeurteilung. Stuttgart (Franckh-Kosmos).

[12] Hemmer, H. & Jaeger, R. (1969): Über ein Pferdeskelett aus dem fränkischen Gräberfeld bei Eltville (Rheingau) nebst Bemerkungen zur Abstammung der Hauspferde. Zeitschrift für Tierzüchtung und Züchtungsbiologie, 85, 3: 221-244.

Figures 2.5-2.6: The second alternative: Exmoor ponies are surviving wild horses and therefore have the same status as Przewalski's horses. Exmoor pony in Exmoor National Park (above, photo Wolfgang Frey) and pure-bred Dzungarian wild horse (Przewalski's wild horse) in Hellabrunn zoo, Munich (below) in comparison.

In 1970, the Exmoor pony was once again recognised as a very old breed of domestic horse, whose origins were though to lie in the earliest Bronze Age horses of Western Europe.[13] From the 1990s

---

[13] Dent, A. A. (1970): The pure bred Exmoor Pony. Williton (Cox).

onwards, the pendulum swung back in favour of the wild horse. In 1998, however, the decision between the two alternatives was left open, for the first time with the help of molecular genetic findings.[14]

In 1999, the Exmoor pony was diagnosed as a relict form of European wild horse, for which a certain influence of domestication cannot be excluded.[15] In view of still unresolved questions, the decision between the two alternatives was initially kept in a certain limbo in 2008 (largely following a first publication in 1993), but with the reference to the conservation breeding of the Przewalski's horse, which is classified as similar to the Exmoor pony, the decision was ultimately made in favour of the wild horse character of the Exmoor pony.[16]

Finally, in 2013, the mandibular and dental studies of the hippological horse type approach were taken up again with supplementary investigations and combined with the current knowledge of mitochondrial DNA, with the conclusion that the Exmoor pony was not a domestic horse, but a wild horse that had remained free of domestic horse influence.[17] In 2018, a skull study, comparing all current equine species once again categorised the Exmoor pony as a domestic breed. In 2022, a review publication on the state of knowledge about horse domestication also cited it as an archaic breed.[18]

All these previous assessments of the nature of the Exmoor pony, which have repeatedly caused the pendulum to swing back and forth between the idea of a breed of domestic horse that has remained original and the interpretation of survival of a wild horse that has been

---

[14] Baker, S., Creig, C.,Macgregor, H., & Swan, A. (1998): Exmoor Ponies – Britain's prehistoric wild horses? Brit. Wildlife, 9: 304-313.

[15] Willman, R. (1999): Das Exmoor-Pferd: eines der ursprünglichsten halbwilden Pferde der Welt. Natur und Museum, 129, 12: 389-407.

[16] Baker, S. (2008): Exmoor-Ponies – Survival of the Fittest – A Natural History. Revised and updated edition. Chippenham (Exmoor Books). First edition 1993.

[17] Hovens, J. P. M. & Rijkers, A. J. M. (2013) : On the origins of the Exmoor pony: did the wild horse survive in Britain? Lutra, 56, 2: 129-136.

[18] Heck, L., Wilson, L. A. B., Evin, A., Stange, M., & Sánchez-Villagra, M. R. (2018): Shape variation and modularity of skull and teeth in domesticated horses and wild equids. Frontiers in Zoology, 15: 14. https://doi.org/10.1186/s12983-018-0258-9

Kyselý, R. & Lubomír Peške, L. (2022): New discoveries change existing views on the domestication of the horse and specify its role in human prehistory and history – a review. Archeologické rozhledy, 74: 299-345.

dominated by humans but has remained largely unchanged over the last two hundred years, are based initially on morphological and later on molecular genetic findings. Both approaches say a lot about the continuity of the populations, but only rarely allow conclusive findings about a domestication that has or has not taken place in the course of this continuity. In the end, the controversy hardly seems resolvable in this way.

In order to understand the problem behind this statement, which relativises previous results, it is necessary to first take a look at the domestication process in general in the following chapter.

Figure 2.7: Exmoor mare with foal in Exmoor National Park. (Photo Wolfgang Frey)

3 Domestication – livestock genesis

3.1 Definition of terms

Domestication is an established term with a long tradition in zoological science. It refers to the process by which wild animals become domestic animals. In the scientific sense, domestic animals are not simply animals that are found in or the immediate vicinity of human dwellings or are kept there for some reason. Domestic animals are forms that can be traced back to wild ancestral species that have been bred for any direct benefit to humans and are clearly different from their wild ancestors. The behaviour of domestic animals appears more subdued than that of wild animals and is determined by the lesser significance of environmental influences. It has been interpreted as the result of a *decline of environmental appreciation.*[19, 20]

In terms of evolutionary biology, a domestic animal is created by the transformation of a wild animal from its perfect behavioural adaptation to its natural environment to a new, equally perfect behavioural adaptation to a different environment established and controlled by humans. It is not created by taming or hand-rearing some individuals, but exclusively by genetic change that intervenes in the nature of the animal. This does not require any targeted human intervention in terms of breeding if sufficient time is available and there is sustained selection pressure.[21]

Outside of zoology, the concept of domestication, which is only clearly defined within this discipline, is watered down beyond recognition. The term domestication is universal, not language-specific and applies to all approaches to keeping usually wild living animals in captivity for the purpose of their utilization. This is domestication in the broadest sense. In this context, domestication is also used as a synonym for taming. However, taming is a reversible change in the be-

---

[19] Hemmer, H. (1990): Domestication – The decline of environmental appreciation. Cambridge (Cambridge University Press).

[20] Several paragraphs of this and the following subchapter are largely taken verbatim from: Hemmer, H. & Carrasco, A. (2023): Erschaffung vielversprechenden Viehs für Lateinamerika. Norderstedt (BoD).

[21] Hemmer, H. (2023): Neumühle-Riswicker – Viehwerdung im Zeitraffer. Norderstedt (BoD).

haviour of individuals, whereas a hereditary change is irreversible if it is not bred out again. Feral domestic animals remain domestic animals from a genetic point of view, even in the wild. If they are reintroduced into captivity, they are easier to tame than their wild ancestors. Domestication in the narrow sense and domestication in the broad sense therefore describe two very different situations. Labelling them with the same name completely obscures this difference.

The power of the factual therefore suggests that it would be better to dispense with this now hollowed-out and widely overstretched term altogether. In this book, therefore, instead of the zoologically unambiguous but otherwise blurred and ambiguous concept of domestication, which is no longer really target-orientated, the origin of domesticated farm animals, including the horse, will be referred to as *livestock genesis*.

When the term livestock is used, it primarily refers to the classic domestic animals that were created and used as farm animals many thousands of years ago. The use of this term in the Holy Books of the world religions of Judaism, Christianity and Islam clearly distinguishes such animals from game. In their essential otherness, livestock were seen as a separate, different branch of creation from wild animals, given to humans for their lasting benefit, as a gift from God. The Bible speaks of this, as does the Qur'an. The bible differentiates "God made the beasts of the earth according to their kinds and the livestock according to their kinds, and everything that creeps on the ground according to its kind".[22] The Qur'an, on the other hand, specifies "Have they not seen that We created for them, of Our Handiwork, livestock that they own? And We subdued them for them [alternative translation *We made them docile*]. Some they ride, and some they eat. And they have in them other benefits, and drinks."[23]

The Qur'an thus refers unmistakably and, differently than with the game, clearly delineates the typical nature of livestock, which is called subdued or docile. The Qur'an, which was written in a society that was crucially dependent on the keeping and everyday use of livestock, correctly formulates the nature of livestock long before its

---

[22] Genesis 1,25 https://www.biblegateway.com/passage/?search=Genesis%201&version=ESV
[23] Sura 36,71-72 https://m.clearquran.com/036.html

scientific consideration. In modern zoology, there is hardly a better characterisation of livestock and certainly not a more generally comprehensible one. Just as the climate is expressed in the weather, the nature of an animal is expressed in its behaviour. Livestock behave differently from game. The term domestic animal in its ambiguity of use, does not contain this clarity. If a successful or unsuccessful livestock genesis is to be assessed, it is first and foremost a question of behavioural characteristics.

Inset 3.1: Duration of classical, evolutionary livestock genesis

For a well-founded idea of the duration of the evolutionary process of livestock genesis, which was completed in early times in adaptation to man-made animal husbandry systems, the beginning and completion of this process would have to be known. As a rule, this is not the case. Archaeology can more or less pinpoint the start of the process with evidence of the first game farming based on various clues. To identify the time of completion, i.e. to prove the immaterial change in nature from game to livestock, will still require a considerable expansion of molecular genetic findings from bone material recovered from archaeological excavations.

The sheep seems to stand for a particularly long period of livestock genesis. Wild sheep farming began around 10,500 years ago in south-west Asia. More than 3,000 years later, animals were shipped from there to Sardinia and Corsica. They are the basic population of all mouflons living in Europe today. According to their behaviour, they are wild sheep and have nothing to do with docile livestock. Nevertheless, they show a certain reduction in brain size compared to their parent group in Southwest Asian mouflons which tends towards domestic animals. It can therefore be assumed that they originate from the beginning, but after more than 3,000 years still not completed livestock genesis period of the sheep, before they became wild again in Sardinia and Corsica.

The rabbit had a particularly short livestock genesis period as a micro-livestock, although its significantly shorter generation time must of course be taken into account. Originally Iberian wild rabbits were kept in the Roman empire from the first century BC. In order to have them quickly available for food purposes, rabbits were subsequently bred in Christian monasteries in the early Middle Ages, where their livestock genesis succeeded in the further course of the Middle Ages. Domestic rabbits can only be said to have existed from the 16th century onwards. An estimate for at least two thousand years for the evolutionary development of larger livestock is therefore by no means too high.[24]

At the end of the previous chapter it was noted that so far only morphological and later additional molecular genetic findings have been collected on the question of the identity of the Exmoor pony. There is a lack of knowledge about the essential difference between the nature of wild horses and domestic horses when it comes to assessing their position in the course if livestock genesis. Although there are extensive behavioural studies on wild living Exmoor ponies, both qualitative and quantitative, there are no quantitative comparative observations of other horses using the same methodology.[25,26] However differences in the nature of the animals as expressed in their behaviour are quantitative differences.

Inset 3.2: Periodic capture and utilization of game does not lead to livestock genesis

It is in the farmer's economic interest to handle the ponies living freely on Exmoor by regularly rounding up the horses in order to skim off the offspring for sale and at the same time regulate the stock with regard to the carrying capacity of the available land. As similar things are done in many places with various livestock that graze freely for long periods of the year on barely limited areas, this fact alone may suggest livestock character of the Exmoor pony.

However, such periodically repeated herding of entire herds of game into trapping facilities for the purpose of obtaining any kind of

---

[24] Almost unchanged from: Hemmer, H. (2023): Neumühle-Riswicker – Viehwerdung im Zeitraffer. Norderstedt (BoD).

[25] Baker, S. (2008): Exmoor-Ponies – Survival of the Fittest – A Natural History. Revised and updated edition. Chippenham (Exmoor Books).

[26] Rödde, S. M.-C. (2015): Verhalten und Raumnutzung von Exmoorponys im Reiherbachtal (Solling). PhD thesis, Georg August-University Göttingen.

product, including stock regulation, does not lead to livestock genesis and is therefore not an indication of it, even if this custom has been practiced for centuries. The start of livestock genesis can only occur if young animals are retained and tamed in such actions in order to build up very long-lasting captive breeding under constricted conditions.

Vicuñas in the highlands of Peru, in the landscape in which the most extensive round-up of game was periodically carried out in Inca times for product extraction and stock regulation, without any livestock genesis process. (Photo Wolfgang Frey)

Perhaps the best example of this can be found in the handling of the vicuña /*Lama vicugna*) during Inca rule. This was probably the most extensive periodic round-up of game in history. Every three to five years, by order of the ruler, a large contingent of 20,000-30,000 people first had the task if erecting a huge enclosure two to three kilometres in diameter with an open entrance about 120 metres wide made of man-high posts and ropes connecting them, on which colourful ribbons fluttered. Then a semicircle formed by all these helpers drove all the game from an area up to 30 kilometres wide towards the gate. Any pumas, bears and foxes caught were usually killed there. Vicuñas and guanacos were tied up using bolas and shorn, while old animals were slaughtered to obtain the meat for drying and the skins. Younger mares and a sufficient number of stallions released again. Apart from shearing the same was done with deer that had also been caught. In this way, population regulation was combined with extensive product extraction. The vicuña wool was reserved for the production of clothing for the Inca upper class, while the wool from the guanacos went to the population. A total of over 20,000 animals are said to have been caught in each such operation.[27] These large-scale harvesting events had no influence whatsoever on the wild character of the vicuña and guanaco. The alpaca and llama had evolved from them thousands of years earlier.

Inset 3.3: Even after centuries of handling, the keeping of game does not automatically result in livestock

Exmoor ponies develop into very good riding and driving horses when they are separated from their wild herd at foal age and subjected to intensive handling by their owner. This is certainly one of the main reasons why they are traditionally seen as one of many breeds of domestic horse. Only if the same can be proven for undisputed wild animals does the alternative question of game or livestock seem worthy of further discussion.

The red deer *(Cervus elaphus)* is a very well-known and widespread large game species for which any though of domestication, of livestock genesis, seems absurd. This is precisely why it is of parallel interest not only for the current Exmoor pony issue. In general, it is a good example of the fact that, on closer inspection, some of the findings that archaeozoologists consider to be indicators of livestock genesis can reveal a great deal about the existence of animal husbandry and diverse human-animal interactions, but become suspect and lose their value as evidence for livestock genesis.

Intense handling of captive red dear has a centuries-old tradition in Europe. The Saxon hunting lodge Moritzburg built at the time of August der Starke (1670-1733) may serve as an early point of reference. The fact that this castle is home to a large collection of particularly strong and conspicuous antlers documents the custom of

---

[27] Gilmore, R. M. (1963): Fauna and ethnozoology of South America. In Steward, J. H. (Ed.): Handbook of South American Indians, 6: 345-464. New York.

mass shooting of deer for this purpose by courtly society in large walled enclosures like no other place. At that time, the breeding of white red deer was also very popular there. Such stags were tamed, given halters and harnesses and used for riding and pulling carts as a curiosity never seen before.[28]

As in the Saxon court three centuries ago, tame red deer led by a halter could be seen in the Popielno Zoological Research Centre (Poland) in more recent times.

Today, red deer play an important role worldwide in agricultural game farming for venison, but also for antler velvet production for traditional East Asian medicine. In the last half century, research-supported husbandry centres have been established in Scotland, New Zealand and Kazakhstan, among other places. Stable husbandry of red deer, which are susceptible and require a high level of veterinary care, has even been practiced in Scotland. However, several centuries of husbandry, breeding and handling of this species in Europe did not result in livestock genesis, but appears to be far from it.

As with red deer, a similar conclusion can be drawn from historical times for other large game species. First and foremost, the two large desert antelopes addax (*Addax nasomaculatus*) and oryx (*Oryx gazella dammah*) should be mentioned for the Old Kingdom of Egypt, the herding of which is documented on the walls of tombs. Before the livestock genesis of horses, hemiones (*Equus hemionus*)

---

[28] Museum Schloss Moritzburg (1993): Die Geweihsammlung Augusts des Starken im Schloß Moritzburg. Moritzburg.

were used in Mesopotamia in the same epoch. Asian elephants (*Elephas maximus*) also began to be used for labour at this time.[29]

In all these cases, the initiation of livestock genesis by keeping and use in captivity obviously did not last long enough to recognize a lasting success in changing the nature of the animals and thus to allow further livestock genesis in an evolutionary way in adaptation to a new, human-dominated environment.

3.2 Preadaptations

Preadaptations play an important role in the evolution of organisms. This term refers to characteristics that have no significant function in the ancestral environment but suddenly become important when the habitat of a species expands into a different environment in which they favour rapid selective adaptation. In the transition from the original environment of a wild species to an altered environment imposed by humans as a result of animal husbandry, preadaptations may have the greatest potential.

One such preadaptation in mammals can be the colour of their coat. In addition to the normal colouring, which can be named "wild colour", deviating colour mutants occur more or less frequently in many species. To name a few well-known examples, black individuals may be considered first, which exist in varying proportions in several populations of leopards ("black panthers") and jaguars, for example. In some areas of Germany, black roe deer occur alongside red roe deer, and black squirrels also occur alongside red squirrels. In the case of roe deer, even piebalds are found, albeit rarely. Perhaps the largest colour spectrum is known from the North American black bear, which also exists as the red "cinnamon bear", the bluish-grey "glacier bear" and the white "ghost bear". None of these species are suspected of ever having had anything to do with domestication. Such colour variants can be enormously important preadaptations for the transition from the natural environment of a species to the man-made husbandry environment.

---

[29] Hemmer, H. (1990): Domestication – The decline of environmental appreciation. Cambridge (Cambridge University Press).

From time immemorial, folk-wisdom has told of different characteristics associated with hair colour in humans and coat colour in some animals. In the 20th century, which felt enlightened and scientific, such views were dismissed as old wives' tales that people felt superior to. Nevertheless, such classifications remained in the minds of equine experts for a long time, who reported in easily remembered rhyming form that black horses were temperamental, brown horses reliable and chestnut horses were more prone to unexpected behaviour. From the middle of the 20th century onwards, more and more evidence of the actual existence if such psychophysical correlations was published. They proved across various mammalian orders that different coat colours on the one hand are accompanied by differences in behaviour on the other. Differences also exist for growth and are primarily found in the stress physiological system, They prove to be highly significant for livestock genesis.[30]

Figure 3.1: Colour mutants as potential preadaptations are sometimes extremely rare, such as this barely known conspicuous colouring of a steppe zebra mare whose foal bears the stripe pattern typical of the population in question (Etosha National Park, Namibia). The coupling of such mutants with changes in temperament in the preadaptation sense has not yet been studied in most cases.

---

[30] Detailed summary in: Hemmer, H. (1990): Domestication – The decline of environmental appreciation. Cambridge (Cambridge University Press).
Hemmer, H. (2023): Neumühle-Riswicker – Viehwerdung im Zeitraffer. Norderstedt (BoD).

Figure 3.2: Sometimes changes in temperament associated with coat colour mutants can already be recognized in photos: in early spring a team member is feeding in a fallow deer enclosure where public feeding is planned and the animals are just developing food tameness again as every spring with increasing number of visitors. The herd includes a majority of wild-coloured deer, menil deer (strong lightening, light belly) and black animals. Menil animals come to feed first, black ones stay to the left of the person immediately next to them, wild-coloured ones remain at a greater distance.

Inset 3.4: Preadaptations – colour mutants in horses before their livestock genesis

According to molecular genetic studies, coat colouring deviating from the wild colouring (colour description in chapter 1) apparently only rarely occurred in Pleistocene wild horse populations. These include the "leopard spotting". There is evidence of black horses on the Iberian Peninsula in the post-glacial period, which became more common during the Neolithic period. The chestnut colour allele was first discovered in Romania in the late 5th millennium BC. After beginning of horse domestication, it was represented in the Bronze Age with a frequency of 28 %. Other colour mutants also increased significantly during the successful livestock genesis.[31] Before that, around 5,500

---

[31] Ludwig, A., Pruvost, M., Reissmann, M., Benecke, M., Brockmann, G. A., Castaños, P., Cieslak, M., Lippold, S., Llorente, L., Malaspinas, A.-S., Slatkin, M., & Hofreiter, M. (2009): Coat Colour Variation at the Beginning of Horse Domestication. Science, 324: 485.

years ago, in the intensively used North Kazakh population, the leopard spotting was found in animals possibly kept in captivity.[32]

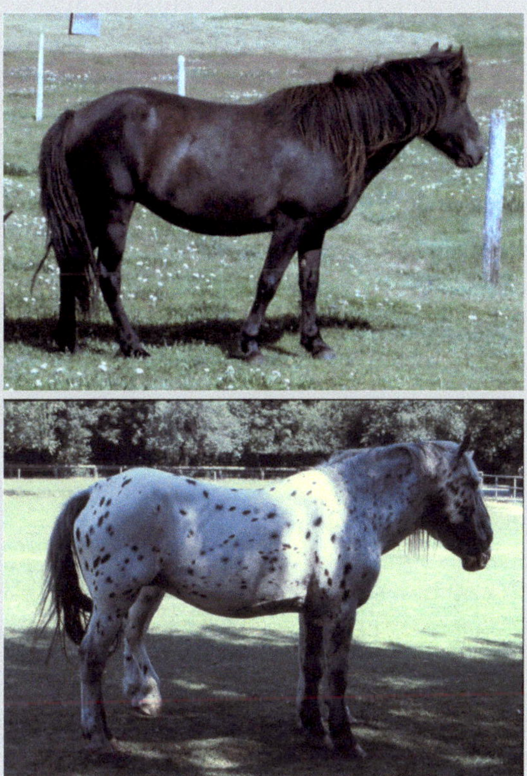

The colour mutants black (here in an Icelandic horse) and leopard spotting (here in a Noriker), potential preadaptations, have been proven by molecular genetics to have existed long before the livestock genesis.

Pruvost, M., Bellone, R., Benecke, N., Sandoval-Castellanos, E., Cieslak, M., Kuznetsova, T., Morales-Muñiz, A., O'Connor, T., Reissmann, M., Hofreiter, M., & Ludwig, A. (1911): Genotypes of predomestic horses match phenotypes painted in Paleolithic works of cave art. PNAS, 108, 46: 18626-18630.

[32] Gaunitz, C., Fages, A., Hanghøj, K., Albrechtsen, A., Khan, N., Schubert,M., Seguin-Orlando, A., Owens, I. J., Felkel, S., Olivier Bignon-Lau, O., et al. & Orlando, L. (2018): Ancient genomes revisit the ancestry of domestic and Przewalski's horses. Science, 360: 111-114.

Another preadaptation for domestication, whose specific function in this context is still not satisfactorily understood, concerns relative brain size. It is almost universal as a criterion of domestication in the proper zoological sense, including the suitability for new domestications of wild species. The term relative brain size refers to the brain mass in relation to the body mass of an animal, the best reference measure for its general body size. Adult large individuals of a population naturally have larger brains on an absolute scale than adult small ones.

However, the relationship between brain mass and body mass is not an isometric one, which would assume the same value on average across all size gradations without changing the proportions. Rather, brain size has a negative allometric relationship to body size. This means that the ratio of both measures also changes with a change in size. With an increase in body mass, the brain mass increases in absolute terms, but decreases in relative terms.

Large, fully grown individuals of the same population therefore have relatively smaller brains than smaller, also fully grown animals. This is expressed mathematically using the so-called allometric formula $BrM = b \times BoM^a$, where BrM is the brain mass, BoM is the body mass, a is the allometric exponent and b is a population-specific integration constant. In mammals, the exponent a fluctuates in the range around 0.23. The constant b is often called the cephalisation constant or cephalisation value in the case of brain size reference and indicates the specific brain developmental stage (cephalisation).[33, 34]

---

[33] Hemmer, H. (1990): Domestication – The decline of environmental appreciation. Cambridge (Cambridge University Press).

[34] Four different methods are primarily used to determine relative brain sizes. In addition to the cephalisation value, these are the extra neuron count, the encephalisation quotient and the progression index. The first two and the last two correlate very closely with each other in terms of their results. The latter are useless for questions of domestication, as they cannot be used to compare closely related populations; they refer to interspecific evolutionary changes within extensive species groups. For a discussion of these methods see Hemmer, H. (2007): Estimation of basic life history data of fossil hominoids. In Henke, W., Tattersall, I. (Eds.): Handbook of Paleoanthropology, Vol. 1: 587-619. Berlin/Heidelberg /New York (Springer).

Inset 3.5: Preadaptations – brain size in horses

Comparative brain size determinations were carried out using two different methodological approaches. The first approach is based on the direct determination of brain and body mass, the second on the measurement of brain volume (cranial capacity) in relation to a length measurement of the skull, the size of which in this case represents body size.

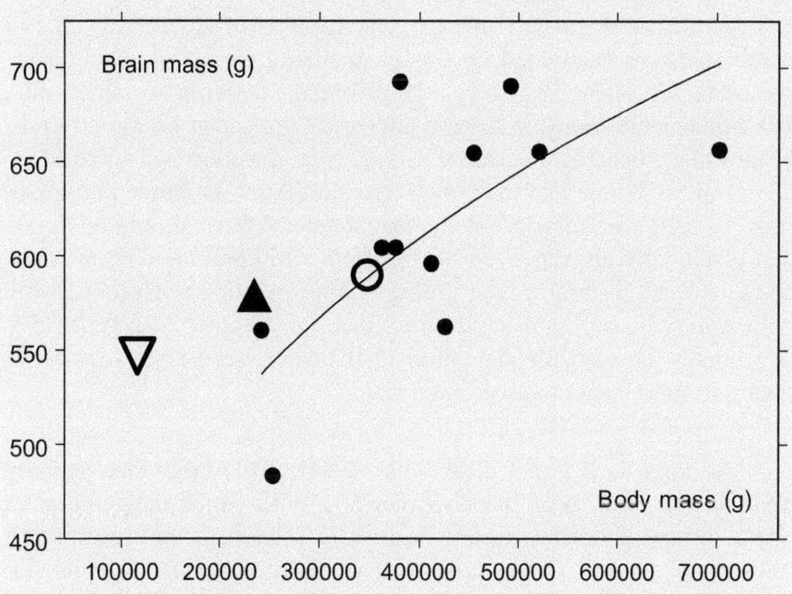

Relation of the brain mass of horses to their body mass (in g). ● *various domestic horse breeds, individual values or mean values for n >1; allometric curve:* brain mass = 23.76 x body mass $^{0.25}$*;* ○ *mean value of domestic horses,* ▼ *mean value of zoo Przewalski's horses,* Δ *mean value of wild Przewalski's horses - not really comparable, as it is mainly based on foal values and is therefore too high. In addition, this value was calculated from skull measurements and not, like the other two means, measured directly.*[37,35]

[35] Crile, G. & Quiring, D. P. (1940): A Record of the Body Weight and Certain Organ and Gland Weights in 3690 Animals. Ohio Journal of Science, 40: 219-249.

In mammals, brain mass grows much faster than body mass during juvenile development. Therefore, adult brain sizes are already reached when the animals are not fully grown.[36] Relating such values to adult values leads to incorrect results, i.e. to an overestimation of the cephalisation level in question. This error occurred when determining the difference in brain size between wild Dzungarian horses (Przewalski's horses) and domestic horses.[37]

The original Dzungerian wild horse had a larger brain than the average domestic horse. However, whether the difference was on average 16 %, as calculated on this problematic methodological basis, should be left open. The very small difference between zoo Przewalski's horses and domestic horses[38] does not appear to be directly relevant either, because the series of values in question is a mixture of horses of the A line and the B line based on domestic horse introgression (for this particular problem, see Chapter 4.5). It should be noted that genetically uninfluenced Dzungarian wild horses had on average up to 16 % larger brains than average domestic horses. In comparison with early wild horses of western Eurasia (cf. Chapter 7.2), Przewalski's horses have no preadaptation to livestock genesis, at least with regard to their larger relative brain size.

As long as it is not clear what specific role brain size actually plays in the process of livestock genesis, with which behavioural or physiological characteristics it is directly related, it may be more a "substrate" for the "impoverishment of the environmental appreciation" rather than as a specific factor for a certain change in behaviour. In evolutionary terms, it seems to precede the change in temperament towards docile livestock. Findings on the European wild sheep, the mouflon, point in this direction. They originally arrived in Sardinia and Corsica, from where mouflons were once introduced to other European regions as game for hunting, in the Neolithic age at the hands of humans, i.e. as farmed game. These European mouflons still behave

---

[36] Hemmer, H. (2023): Neumühle-Riswicker – Viehwerdung im Zeitraffer. Norderstedt (BoD).

[37] Gorgas, M. (1966): Betrachtung zur Hirnschädelkapazität zentralasiatischer Wildsäugetiere und ihrer Hausformen. Zoologischer Anzeiger, 176: 227-235.

[38] Röhrs, M. & Ebinger, P. (1998): Sind Zooprzewalskipferde Hauspferde? Berliner Münchner Tierärztliche Wochenschrift, 111; 273-280.

like wild sheep, but have a slightly reduced brain size compared to the Asian source populations.[39]

If different populations of a species from which livestock originated have different relative brain sizes, then it is always the population with the lowest cephalisation from which the livestock's ancestors originate. The sheep is one if these species, the pig another.[40] A preadaptation to livestock genesis is therefore based on difference in brain size.

3.3 Systematics, taxonomy and nomenclature

To understand the scientific terms used in connection with livestock genesis, an excursion into the world of zoological systematics, taxonomy and nomenclature based on evolutionary processes is essential. Evolution takes place within the framework of populations. A *population* is the totality of all individuals that are in at least a potential reproductive community. The splitting up of populations in the course of their dispersal or their separation through ecological change creates population relationships, initially between populations that are spatially neighbouring and between which there can be an exchange (gene flow) of genetic material. Together they form a population community, a *metapopulation*.

With increasing hereditary diversity in one or more individual populations of the metapopulation, the emergence of *isolation mechanisms* is possible. Isolation mechanisms make it either more difficult to mate (pre-mating mechanisms) or limit the fertility of their offspring, in extreme cases hindering their viability (post-mating mechanisms). The emergence of isolation mechanisms initiate the permanent separation of individual subpopulations from metapopulations.

This leads to the development of new species. Similar to human family clans, there are natural relationships to ancestors (ancestral species), to siblings (sister species) and to descendants (daughter

---

[39] Pees, W. & Hemmer, H. (1980):Hirngröße und Aktivität bei Wildschafen und Hausschafen (Gattung *Ovis*). Säugetierkundliche Mitteilungen, 28, 1:39-45.

[40] Hemmer, H. (1990): Domestication – The decline of environmental appreciation. Cambridge (Cambridge University Press)

species). As a result, increasingly complex kinship structures develop over the course of time and through dispersal in space. Their study is the task of *systematics*.

In order to catalogue their findings and thus make them clear like a branched family tree, an internationally defined classification procedure is required, the *taxonomy*. Despite all attempts at scientific objectivity, a great deal of subjectivity comes into play here, as the relationships created by the sliding dynamic evolutionary processes now have to be broken down into static units (*taxa*) for the purpose of delimiting labelling – squaring the circle, so to speak. Such units are – among many others – the *family*, the *genus* and the *species*.

There is a huge and central problem at the species level. How to define a species? There are two main competing concepts to consider. The *biological species concept* is in fact based on the existence of isolation mechanisms. It is the behaviour and development-based concept. The *phylogenetic species concept*, on the other hand, is a purely trait-based concept that assumes the existence of consistent differences without having to understand their functional significance and their role in evolutionary processes. Its application is simpler and usually results in a higher number of species, but does not provide any information about the biological, evolutionary significance in space and time. Neither of these concepts is the "right" or the "wrong" one. The "species" is an artificial product, not a truly natural entity in the context of the usual use of the term.[41]

With regard to the horse at the centre of this book, two further intermediate or subcategories of taxonomy must be mentioned. Between genus and species there is the taxon *subgenus*, which is used when a genus contains several species that have split off from one another in more than one evolutionary step. A species can be composed of several *subspecies*. In no case do they fulfill their own species criteria. They are defined in such a way that at least 75 % of the individuals can be assigned to a specific population. Their existence expres-

---

[41] On the species problem and the metapopulation concept:
Queiroz, K. de (2005): Ernst Mayr and the modern concept of species. Proceedings of the National Academy of Sciences of the United States of America 102, supplement 1, 6600–6607.

ses the possible marginal gene exchange between individual populations within a metapopulation.

Taxonomic units have internationally binding names, which are essential for clear understanding. This is the turn of the *nomenclature*. It is regulated in the form of a set of laws. Names of taxa are written in Latin or ancient Greek or, in the case of terms from modern languages, preferably with corresponding endings. The first name given to a particular taxon takes precedence over all subsequent names (principle of priority). Names at the level of at least genera begin with capital letters, names at the species level with lower case letters. Species names consist of two words, namely the genus name plus the species name. Subspecies names contain three words by adding the subspecies name to the species name. Genus, species and subspecies names are written in italics. Subgenus names can be inserted in brackets. The complete spelling also includes the name of the original author and, separated by a comma, the year of publication.

These taxonomic and nomenclatural regulations are presented using the example of the horse. All modern horses are summarized in the genus *Equus* Linnaeus, 1759. This genus is subdivided into several subgenera, whereby only those with living species are listed here:

Equus (*Asinus*)	asses
Equus (*Hemionus*)	hemiones
Equus (*Dolichohippus*)	Grevy's zebras
Equus (*Hippotigris*)	mountain zebras
Equus (*Quagga*)	steppe zebras
Equus (*Equus*) – horses in the narrower sense, caballine horses	

The subgenus *Equus* includes the wild horses of the species *Equus* (*Equus*) *ferus* Boddaert, 1785, and the domestic horse *Equus* (*Equus*) *caballus* Linnaeus, 1758. If the Dzungarian wild horse is treated as the same species as *Equus ferus* using the biological species concept, it receives subspecies status as *Equus ferus przewalskii* Poljakoff, 1881, which is contrasted with the Russian tarpan as *Equus ferus ferus* Boddaert, 1785 (chapter 4.4). The phylogenetic species concept would also allow *Equus przewalskii* Poljakoff, 1881, for the Przewalski's horse.

Figures 3.3-3.8: Wild ass – Equus (Asinus) africanus; domestic donkey – Equus (Asinus) asinus (top); hemione – Equus (Hemionus) hemionus; Grevy's zebra – Equus (Dolichohippus) grevyi (centre); mountain zebra – Equus (Hippotigris) zebra; steppe zebra – Equus (Quagga) quagga (bottom).

The nomenclatural designation of the domestic horse was discussed for decades as part of the question of the naming of domestic animals in general. As domestic animals are descended from wild animals, livestock from game, the argument has repeatedly been in favour of species equality. According to the priority principle of nomenclature, this decision should have the consequence for almost all classic domestic animals – with the exception of pigs – that the wild ancestors should have to bear the name of their domestic descendants. All wild horses would have to go by the name *Equus caballus* as long as they are treated as the same species as the domestic horse. Without being allowed to intervene in the discussion on whether wild and domestic animals are the same species or different species, the International Commission on Zoological Nomenclature finally solved the problem in 2003 with a Solomonic, fundamentally binding decision. Although scientific names given to domestic species remain valid, they lose their original priority over the names of their wild ancestors.[42]

As there are enough factual arguments against the identity of domestic animals with their wild ancestors, this problem is solved anyway. As a rule, there are clear distinguishing features between livestock and game. The separation with the phylogenetic species concept is therefore beyond question. The biological species concept, which refers to isolation mechanisms, can also underline this. Social sorting as a pre-mating isolation mechanism has been documented where it has been studied.[43]

It is perhaps not entirely wrong to speak of a certain chaotic situation in the systematics, taxonomy and nomenclature of horses until the end of the 20th century, whether in zoology or palaeontology. Only with the progress in molecular genetics in the last decade has there been a beginning clarity, as the two following chapters show.

3.4 The decision question

This brings up the controversy surrounding the nature of the Exmoor pony. For the zoological classification with a certain taxon and thus also for its internationally valid scientific naming, an either/or decision is unavoidable. Either the Exmoor pony is to be understood as a

---

[42] Opinion 2027 (case 3010) (2003): Usage of 17 specific names based on wild species which are pre-dated by or contemporary with those based on domestic animals (Lepidoptera, Osteichthyes, Mammalia): conserved. Bulletin of Zoological Nomenclature, 60)1): 81-84.

[43] Hemmer, H. (2015): Neumühle-Riswicker – Viehwerdung im Zeitraffer. Norderstedt (BoD).

wild horse of the species *Equus ferus* or it is a domestic horse of the species *Equus caballus*. This nomenclatural decision has been avoided by all previous studies, no matter how extensive and carefully design-ed. It must be made at the end of this book.

The main issue is therefore whether or not there is evidence of a difference in character between the Exmoor pony and livestock, i.e., domestic horse breeds. This can only be tested with comparative-quantitative behavioural studies. In this sense, the formulation that it is "almost a dispute about words whether the Exmoor horse is called wild horse or not" is too short-sighted.[44] The embarrassment as to what the Exmoor pony really is taxonomically can hardly be better expressed than with the nomenclaturally impossible designation as *Equus feruscaballus* in an ecological publication.[45]

---

[44] Willman, R. (1999): Das Exmoor-Pferd: eines der ursprünglichsten halbwilden Pferde der Welt. Natur und Museum, 129, 12: 389-407.

[45] This species name is not a misprint, as the species name *feruscaballus* is actually weitten together as one word in both the title and the abstract:

Hagstrup M., Bruhn D., Olsen K., Lukassen M., & Pertoldi C. (2020): Molecular study of dietary diversity of the Exmoor-ponies (*Equus feruscaballus*). Genetics and Biodiversity Journal, 4, 1: 53-70.

Figure 3.9: The decision question of zoological taxonomy and nomenclature: are Exmoor ponies domestic horses of the species Equus caballus or are they wild horses of the species Equus ferus? (Photo Wolfgang Frey in the Exmoor National Park)

4 Wild horses

4.1 Knowledge from palaeontology

Real horses of the subgenus *Equus* only became widespread in Eurasia from the early Middle Pleistocene onwards with the evolutionarily archaic taxon *Equus ferus mosbachensis* von Reichenau, 1915. After that, the Middle and Late Pleistocene cyclical change between climatic extremes, between cold and warm periods, resulted in repeated ecosystem changes between tundra, steppe and forest in regionally different ways. Horses populated these very diverse landscapes in large numbers, albeit with functionally differently adapted forms and a wide variety of body sizes.[46]

According to a general zoological climate rule (Allen's rule), the extremities of Pleistocene horses are shorter relative to the body under cold-climate conditions, so that the load-bearing bone elements appear more robust relative to their length due to the same load from the body mass. This is particularly evident in the metapodial bones. Short and broad snouts are seen as a cold-climate adaptation, while elongated and often narrowed snouts are attributed to warm-climate conditions.

---

[46] Pleistocene evolution of the horse in Europe:

Boulbes, N. & van Asperen, E.N. (2019): Biostratigraphy and Palaeoecology of European *Equus*. Frontiers in Ecology and Evolution, 7, 301: 1-30. https://doi: 10.3389/fevo.2019. 00301

Eisenmann, V. (1991): Les chevaux Quaternaires Européens (Mammalia, Perissodactyla). Taille, typologie, biostratigraphie et taxonomie. Geobios, 24, 6: 747-759.

Eisenmannn, V. (2022): Old World Fossil *Equus* (Perissodactyla, Mammalia), Extant Wild Relatives, and Incertae Sedis Forms. Quaternary, 5, 38. https://doi.org/10.3390/ quat5030038

Forsten, A, (1988): The small caballoid horse of the upper Pleistocene and Holocene. Journal Animal Breeding Genetic, 105: 161-176.

Forsten, A. & Ziegler, R. (1995): The horses (Mammalia, Equidae) from the early Wuermian of Villa Seckendorff, Stuttgart-Bad Cannstatt, Germany. Stuttgarter Beiträge zur Naturkunde Ser. B (Geologie und Paläontologie), 224.

Forsten, A. & Dimitrijević, V. (2004): Pleistocene horses (genus *Equus*) in the central Balkans. Annales Géologiques de la Péninsule Balcanique, 65: 55-75.

Guadelli, J.-L. (1991): Les chevaux de Solutré (Saône et Loire, France). Cahiers du Quaternaire, 16 : 261-336.

Nobis, G. (1971): Vom Wildpferd zum Hauspferd – Studien zur Phylogenie pleistozäner Equiden Eurasiens und das Domestikationsproblem unserer Hauspferde. Fundamenta, B, 6. Köln/Wien (Böhlau).

Spassov, N. & Iliev, N. (1997): The wild Horses of Eastern Europe and the polyphyletic origin of the domestic horse. Anthropozoologica, 25-26: 753-761.

The size, position and chewing surface structure of the cheek teeth (premolars and molars) have a decisive influence on the processing of the different food supply in different habitats with different climatic conditions.

As already vividly illustrated for the hippological pony and horse types (horse types according to Speed and Ebhardt) with X-ray images, the grinding teeth are relatively larger when mainly low-energy coarse forage is available. Diagnostic significance in the enamel edge structure of the grinding surfaces of the upper cheek teeth, which is decisive for abrasion, is attributed to the shape and size of the protocone, the inner pillar. Three types (palaeontological horse types according to Eisenmann) are distinguished according to its relative size. Type I horses have on average a shorter protocone of the third and fourth premolars compared to the first and second molars. They are generally found in temperate climates. The majority of Late Pleistocene horses in Europe represent this type. Horses of tooth type II show the opposite pattern. In them, the protocone is long in the last two upper premolars and shorter in the first two molars. This type is found more frequently in cold-climate conditions, but is rare overall. It occurs in small horses living in the latest glacial period, in the west during the Magdalenian culture. In eastern Europe it is found as far as the Desna in the Ukraine. The protocone of tooth type III is short in both premolars and molars.

In terms of hippological horse typology, there were two very different horse forms in the European Late Pleistocene, a heavier-built, broad-hoofed pony type and a lighter-built, finer-boned horse type. It is disputed whether these two forms actually lived side by side in the same area for a longer period of time as some authors assume. If this were the case, their herds would have remained separate without mixing, thus providing evidence for the existing of pre-mating isolation mechanisms, a prerequisite for the emergence of separate species[47]. However, molecular biology analysis of the last Ukrainian wild horses

---

[47] On the concept of isolation mechanisms see chapter 3.3. For the use of such considerations in palaeontology see Hemmer, H. & Kahlke, R.-D. (2022). New results on felids from the Early Pleistocene site of Untermassfeld. In R.-D. Kahlke, (Ed.), The Pleistocene of Untermassfeld near Meiningen (Thüringen, Germany). Part 5. Monographien des Römisch-Germanischen Zentralmuseums Mainz, 40(5), 1465–1566.

confirms broad genetic combination of both forms (see 4.4: The tarpan). Thus, even in view of this enormous diversity, this author considers that they were temporally and spatially different subspecies, not several species.

For the horses of the western European region, a tendency towards a reduction in size can be recognised in the Late Pleistocene, starting from *Equus ferus germanicus* Nehring, 1884, representing tooth type I, to *Equus ferus gallicus* Prat, 1968. now regarded as its smaller late form. Finally, even smaller ponies with tooth type II, *Equus ferus arcelini* Guadelli, 1987, are documented from French deposits dated to the Würm latest glacial period, younger than 15,000 years ago. A taxonomically indisputable separation of all three chronologically consecutive populations at the subspecies level can hardly be assumed for the first two. For the third, it is more acceptable, since both size as well as tooth type changes go hand in hand compared to the previous ones. A Late Pleistocene Iberian population – *Equs ferus antunesi* Cardoso et Eisenmann, 1989 – comes close to *Equus ferus arcelini* with regard to the dentition, but not to the metapodials.

The population group around *Equus ferus germanicus*, which also settled in south-eastern and eastern Europe as far as the Ukraine in the Late Pleistocene, was initially understood there as a separate form, *Equus ferus latipes* Gromova, 1949. The entire group is characterised by broad hooves, robust phalanges, broad metapodials and large cheek teeth with a relatively long protocone. The structural characteristics of these so-called "broad-footed horses" are interpreted as an ecological adaptation to the softer soils of forest-steppes and steppe-tundras. The distribution area of these horses, which primarily represent pony type I in terms of hippological typology, reached as far east as the Don, where they could still have contact after the Ice Age with morphologically contrasting steppe horses, which correspond with horse-type IV in terms of horse typology.

Since the Late Pleistocene, the Russian tarpan, *Equus ferus ferus* Boddaert, 1785, was widespread in the steppe landscape in the extreme east of Europe, finally condemned to extinction during the 19th century. This horse, which was adapted to the hard soils of wide, open steppes, was characterised by narrow hooves and metapodials, as well as smaller cheek teeth with short, wider protocones.

4.2 Knowledge from Archaeozooology

Palaeontological studies of horses up to the end of the Pleistocene focused on evolutionary kinship analyses by means of morphological comparisons. In contrast, for the following millennia, archaeozoological studies focused primarily on the interaction between humans and horses and their domestication. The general problem after the Neolithic is the morphological distinction between wild and domestic horses.

After the end of the last ice age, horses became increasingly rare in most parts of Europe to the point of complete disappearance from the archaeological record. From the time when Europe was largely characterised by forestation during the early Atlantic Period (around 9,100-7,500 years ago), only very few horse remains have been found in the peripheral areas of continental Europe, but these provide fundamental evidence of the continued existence of European relict populations. During this time the previously glacial land connection between the continent and England was flooded and interrupted by the formation of the English Channel as a resettlement barrier. It was only later that horses were again frequently found throughout continental Europe. This was interpreted – albeit not uncontradicted – as the recolonisation of open landscapes resulting from Neolithic agriculture. From England, the region of central interest for this book, the horse has only been proven with radiocarbon dating up to around 10,000 to 9,800 years ago. However, its small-scale survival until its reappearance around 3.700 years ago cannot be ruled out.[48] From this time onwards, there is the problem that horse remains found in the context of settle-

---

[48] Clutton-Brock, J. & Burleigh, R. (1991): The Mandible of a Mesolithic Horse from Seamer Carr, Yorkshire, England. In Meadow, R. H. & Uerpmann, . P. (Eds.): Equids in the Ancient World, Volume 2: 238-241. Wiesbaden (Reichert).
Clutton-Brock, J. & Burleigh, R. (1991): The Skull of a Neolithic Horse from Grime's Graves, Norfolg, England. In Meadow, R. H. & Uerpmann, . P. (Eds.): Equids in the Ancient World, Volume 2: 242-249. Wiesbaden (Reichert).
Kyselý, R. & Lubomír Peške, L. (2022): New discoveries change existing views on the domestication of the horse and specify its role in human prehistory and history – a review. Archeologické rozhledy, 74: 299-345.
Sommer, R. S., Benecke, N., Lõugas, L., Nelle, O., & Schmölcke, U. (2011): Holocene survival of the wild horse in Europe: a matter of open landscape? Journal of Quaternary Science, 26, 8: 805-812

ments in Europe are generally attributed to domestic horses, although the simultaneous existence of wild horses in many regions can still be assumed.

Horses from a northern Kazakh population west of the Dzungarian Gate, which were very close to the current *Equus ferus przewalskii*, were used to a considerable extent in the Botai culture around 5.500-5.000 years ago. With the assumption that this was the first livestock genesis of wild horses, long before the emergence of today's domestic horses, the additional assumption was made that today's Przewalski's horses are not wild horses at all, but feral descendants of such first and independently domesticated horses.[49] Even if the archaeological evidence for a livestock genesis process, which is hardly to be expected in the short period of time in question, were conclusive, this subsequent idea does not necessarily seem worthy of further discussion.[50] In any case, the wild horse character of the Przewalski's horse cannot be called into question with this idea.

4.3 Knowledge from molecular genetics

Molecular genetic findings on the entire genome make it possible to estimate the chronological fitting of population splits as the beginning of separate evolutionary lines. During the late Middle Pleistocene, in the period 335,000-285,000 years ago, a lineage of western European horses first separated ftom other Eurasian populations. Palaeontologically, this branch is possibly the ancestor not only of the European broad-footed horses, hut also of the later Iberian subspecies.

Next, a Siberian population (*Equus ferus lenensis* Russanov, 1968) split off from the rest of the Eurasian wild horses at the begin-

---

[49] Gaunitz, C., Fages, A., Hanghøj, K., Albrechtsen, A., Khan, N., Schubert,M., Seguin-Orlando, A., Owens, I. J., Felkel, S., Olivier Bignon-Lau, O., et al. & Orlando, L. (2018): Ancient genomes revisit the ancestry of domestic and Przewalski's horses. Science, 360: 111-114.

[50] If one were to follow this idea that all modern-day Przewalski's horses are feral domestic horses, because five millennia ago domestic horses evolved from one of their distant sub-populations, then this must also be postulated for other game-livestock lineages. Then all today's guanacos would be feral llamas, all vicuñas would be feral alpacas, all wild yaks would be feral domestic yaks, all wild water buffalo would be feral domestic buffalo, not to mention the wild boar.

ning of the Late Pleistocene. The separation of the Eastern European-Central Asian steppe belt populations leading to the Przewalski's horse and the ancestors of the domestic horse only occurred 55,000-35,000 years ago, i.e. during the Late Pleistocene in the first half of the Würm Glacial, whereby the distribution area of the early Przewalski's horses extended westwards to the Urals. A rapid and wide spread of horses from the steppes around the lower Don and the lower Volga, which correspond to modern domestic horses in terms of molecular genetics, only began with the progress of their domestication around 4,200 years ago. Before that, such horses can only occasionally be traced further west to the lower Danube region of Romania.[51]

During Pleistocene warm phases, repeated collapse and fragmentation of horse populations is indicated by molecular genetics, very clearly for the last interglacial.[52] The recolonisation of ecologically changing landscapes from relic populations that solidified during cold periods created the basis for the palaeontological diversity of body size, shape and structure of the dentition.

4.4 The tarpan

The first scientific observations and descriptions of still common wild horses in the southern Russian forest-steppe and steppe regions around the Don and Volga rivers, called tarpans by the local Tatar population, led at the end of the 18th century to the zoological naming *Equus ferus* Boddaert, 1785, for wild horses from the Russian region of Bobrov southeast of Voronezh.

At this time, the international rules of zoological nomenclature did not yet require a defined specimen (type specimen procedure) for a newly introduced taxon in order for the name to be valid. This had serious consequences for the later understanding of the identity of the

---

[51] Orlando, L.(2019): Ancient Genomes Reveal Unexpected Horse Domestication and Management Dynamics. BioEssays, 42, 1. https://doi.org/10.1002/bies.201900164
Librado, P., Khan, N., Fages, A., Kusliy, M. A., Suchan, T., et al. & Orlando, L. (2021): The origins and spread of domestic horses from the Western Eurasian steppes. Nature, 598. https://doi.org/10.1038/s41586-021-04018-9
[52] Orlando, L., Ginolhac, A., Zhang, G., Froese, D., Albrechtsen, A., Stiller, M., et al. (2013): Recalibrating *Equus* evolution using the genome sequence of an Early Middle Pleistocene horse. Nature, 499: 74-78.

tarpan. The pure-bloodedness of the few skeletal remains of three individuals from the time of its extinction in late 19th century has been called into question. They do not originate from the Russian Don-Volga region, but from the southern Ukrainian steppe around the lower reaches of the Dnipro in the Kherson region.

With the later discovery of further modern and Pleistocene wild horse forms, which were ultimately categorized as being from the same species but with subspecies differences, the Russian tarpan population had to be automatically elevated to the rank of the nominate subspecies *Equus ferus ferus*.[53] The southern forest-steppe and the steppe around the Don and Volga rivers from which it was described, is the molecular-genetically derived area of origin of horse domestication, so that the wild horse branch on which it is based is clearly recognisable as the tarpan *Equus ferus ferus*.

The little bone material still available was considered morphologically typical of the tarpan in comparison with finds from the time before domestication.[54] The proponents of the hippological horse typology had difficulty in categorising this last piece of evidence because, according to their findings, it fits neither one of the pony types nor one of the horse types, but rather occupies an intermediate position with the combination of characteristics of both lines.[55]

Remarkably, this result is fully supported by molecular genetics.[56] About one third of the genome of the last southern Ukrainian wild horses considered to be tarpans can be assigned to western

---

[53] The new description of an *Equus ferus ferus* for a late Ice Age broad-hooved horse of this region is nomenclaturally invalid due to this automatism:
Nobis, G. (1971): Vom Wildpferd zum Hauspferd – Studien zur Phylogenie pleistozäner Equiden Eurasiens und das Domestikationsproblem unserer Hauspferde. Fundamenta, B, 6. Köln/Wien (Böhlau).
[54] Spassov, N. & Iliev, N. (1997): The wild Horses of Eastern Europe and the polyphyletic origin of the domestic horse. Anthropozoologica, 25-26: 753-761.
Eisenmannn, V. (2022): Old World Fossil *Equus* (Perissodactyla, Mammalia), Extant Wild Relatives, and Incertae Sedis Forms. Quaternary, 5, 38. https://doi.org/10.3390/ quat5030038
[55] Speed, J. G. & Etherington, M. G. (1952): Am aspect of the evolution of British horses. The British Veterinary Journal, 108. 5. Reprint in Speed, J. G. & Speed, M. G. (Eds./1977): The Exmoor-Pony, 14-25. Chippenham (Countrywide Livestock Ltd).
Schäfer, M. (2000): Handbuch Pferdebeurteilung. Stuttgart (Franckh-Kosmos).
[56] Librado, P., Khan, N., Fages, A., Kusliy, M. A., Suchan, T., et al. & Orlando, L. (2021): The origins and spread of domestic horses from the Western Eurasian steppes. Nature, 598. https://doi.org/10.1038/s41586-021-04018-9

European horses, two thirds are close to the domestic horse ancestors of the Don-Volga region. As the latter are the nomenclaturally true tarpans, the museum specimens obviously represent a wild horse population that appears to have originated from the mixing of *Equus ferus ferus* steppe horses with broad-footed horses from the west. This finding on the last Ukrainian wild horses classically attributed to the tarpan, which only appears surprising at the first glance, is not really surprising, but rather reflects the historical potential for mixing that is to be expected from the at least temporarily interlinked distribution of the western European and eastern subspecies, of populations not separated on the species level.

The southern Ukrainian wild horses that emerged from significant introgression of broad-footed horses into an *Equus ferus ferus* population were therefore neither European broad-footed horses nor Russian tarpans. Due to their similarity to the tarpan, their predominant genome affiliation and their traditional assignment to the tarpan, they may be called Ukrainian tarpan in contrast to the Russian tarpan. In taxonomic terms, the name *Equus ferus gmelini* Antonius, 1912, is available for the Ukrainian tarpan. The former Polish wild horses known as forest tarpans (*Equus ferus silvestris* Brinken, 1826), about which very little of substance is known, may have been horses with an even higher proportion of broad-footed horses.[57] Drawings of Russian tarpans by the first scientific observers of the 18th century[58] show horses that differ from the only existing photo (1884) of an adult Ukrainian tarpan in their more graceful shape with relatively longer legs, similar to oriental horses.

4.5 The Przewalski's horse

In Inner Asia, the Dzungarian wild horse, the Przewalski's horse (*Equus ferus przewalskii* Poljakoff, 1881), named after its scientific discoverer, survived in the border region between western Mongolia and Xinjiang in north-west China until the 1970s. Conservation

---

[57] On the synonymy of Russian-Ukrainian wild horses and their former distribution see Heptner, V. G., Nasimovich, A. A., & Bannikov, A. G. (1988): Mammals of the Soviet Union, Vol. 1, Artiodactyla and Perissodactyla. Washington (Smithsonian).
[58] https://de.wikipedia.org/wiki/Tarpan

breeding in captivity has secured its continued existence for the time being.

Figure 4.1: Dzungarian wild horse or Przewalski's horse (Front Royal, Virginia, USA)

However, this poses a problem that cannot be overlooked.[59] The first attempts to capture adult Przewalski's horses at the end of the 19th century, around 20 years after their discovery, failed due to their great shyness and enormous speed. Numerous very young foals that were eventually captured, were characterised by a high mortality rate. As adults, many of the surviving horses which were spread across several zoological gardens, scientific institutions and large private holdings, produced no or very few offspring. As a result of this poor reproductive performance, descendants of a crossbred stallion from a Przewalski's stallion and a Mongolian mare transported to Europe as a nurse mare were integrated in the breeding program.

---

[59] Bouman, I. & Bouman, J. (1994): The History of Przewalski's Horse. In Boyd, L. & Houpt, K. A. (Eds.): Przewalski's Horse – The History and Biology of an Endangered Species: 5-38. New York (State University of New York Press).

This early crossbreeding resulted in easier breedability,[60] which ensured that the breeding line thus established (the so-called B line or Prague line) moved to the forefront in terms of the number of individuals in the entire population. It also meant that Przewalski's horses from zoo breeding showed skeletal morphological differences to the original wild form[61] and that the question was raised as to whether such horses were domestic horses in terms of brain size.[62]

Figure 4.2: Former breeding group of Przewalski's horses of the A line at Hellabrunn Zoo (Munich)

There are still Dzungarian wild horses in the so-called A (or Munich) line with introgression-free ancestry in terms of molecular genetics,[63] the number of which should in principle be sufficient for

[60] Heck, H. (1976): Die Erhaltung des Przewalskipferdes. Offprint of the 3rd International Symposium on the Conservation of the Przewalski's horse, München, 26-28 April 2076.

[61] Forsten, A, (1988): The small caballoid horse of the upper Pleistocene and Holocene. Journal Animal Breeding Genetic, 105: 161-176.

[62] Röhrs, M. & Ebinger, P. (1998): Sind Zooprzewalskipferde Hauspferde? Berliner Münchner Tierärztliche Wochenschrift, 111; 273-280.

[63] Orlando, L., Ginolhac, A., Zhang, G., Froese, D., Albrechtsen, A., Stiller, M., et al. (2013): Recalibrating *Equus* evolution using the genome sequence of an Early Middle Pleistocene horse. Nature, 499: 74-78.

Der Sarkissian, C., Ermini, L., Schubert, M., Yang, M. A., Librado, P., Fumagalli, M., Jónsson, H., Bar-Gal, G. K., Albrechtse Conservation of the Endangered Przewalski's Horse. Current Biology, 25, 19: 2577-2583.

the conservation of pure-bred Przewalski's horses. Although they do not represent the entire genetic spectrum of all founder animals, which is only preserved in the B line, the separate retention of both lines would fulfill all conservation requirements. Horses from the B line in particular were released into the wild in their original range.[64]

A certain introgression of domestic horse genes in the original Dzungarian wild population, which has been discussed time and again,[65] would have been less problematic in terms of its wild or livestock character compared to those finally in Mongolia and Xinjiang especially with B line horses newly established free-range populations. In the former case, immediate selection-pressure against livestock-type traits would have been expected in hybrids; in the latter, long-term selection in the captive environment was in their favour. The fact that this had an effect on the social behaviour of the horses, which is essential for their release into the wild, is shown by a correlation between the social aggressiveness of mares in zoo environment and their calculated domestic horse ratio.[66]

Comparative-quantitative studies on horses reintroduced to their original habitat and associated photographic material show differences in aggressive behaviour between different harem groups, which may be related to this.[67] Nevertheless, a certain effect of selection pressure in natural environment against behaviour typical of domestic horses seems to be emerging.

---

[64] Bouman, I., Bouman J., & Boyd, L. (1994): Reintroduction. In Boyd, L. & Houpt, K. A. (Eds.): Przewalski's Horse – The History and Biology of an Endangered Species: 255-263. New York (State University of New York Press).

[65] Robovský, J. (2009): Przewalski horse: a review of controversies over its raxonomy, phylogeny and full-bloodedness. Equus 2009, Zoo Praha: 57-112.
Zimmermann, W. (2009): The domestic trait(s) in the A-line of PrzewalskiÄs horses (*Equus ferus przewalskii*). Equus 2009, Zoo Praha: 229-255.
The assumption that domestic horse introgressions were already present in the founder animals is based on the original postulate of type purity from a horse-breeding perspective, but not on the variability to de assumed in every natural population. With today's molecular genetic knowledge, this question can be put to rest.

[66] Greif, A. (1989): Beobachtungen an Przewalskipferden (*Equus ferus prewalskii*) im Hinblick auf ihren Hauspferdanteil. Diploma thesis, FB Biology, Johannes Gutenberg-University Mainz.

[67] Hoesli, T., Nikowitz, T., Walzer, C., & Kaczenski, P. (2009): Monitoring of agonistic behaviour and foal mortality in free-ranging Przewalski horse harems in the Mongolian Gobi. Equus 2009, Zoo Praha: 113-135.

Figure 4.3: Is the Exmoor pony a wild horse like the Przewalski's horse and the tarpan? (Photo Wolfgang Frey in the Exmoor National Park)

5 Domestic horses

5.1 History

According to uncertain archaeological evidence, broad-footed horses may have been tried to manage in captivity in Eastern Europe more than 6,000 years ago. In the period around 5,500-5,000 years ago, wild horses from a North Kazakh population very similar to today's Przewalski's horses may also have been kept in captivity (Chapter 4.2). However they all have little to do with the livestock genesis of extant domestic horses.[68]

After possibly two millennia of little success to keep wild horses in various areas of Europe and Inner Asia, this took off around 4,200 years ago in the population of the lower Don-Volga region and led almost explosively to the spread of early domestic horses across Europe and Asia within a few centuries until the beginning of the second millennium BC. In both continents, they came into contact with regional wild horse populations, from which they ultimately received small amounts of genetic material. The actual livestock character of these first domestic horses is confirmed by evidence of selection of a docility gene on the neurotransmitter pathway and thus the change in character postulated for livestock.

Two genetically separated lines of development emerged on the way to today's European breeds. On the one hand, there are Celtic, Gallo-Roman, Piktish and Viking horses, including Shetland and Icelandic ponies. This line may be called the north-west European, Celtic-Germanic breed group. On the other side are mainly continental European breeds, Friesians, Dülmeners, Sorraias, Connemaras, as well as modern warm-blooded and thorough-bred breeds. The differentiation of this group is understood to have come about as a result of new horse imports in the course of early Medieval Arab colonisation in areas bordering the Mediterranean. This group of breeds may therefore be called the oriental one. From this time onwards, there was a growing influence of Oriental bloodlines on European horse breeding.

---

[68] Kyselý, R. & Peške, L. (2022): New discoveries change existing views on the domestication of the horse and specify its role in human prehistory and history – a review. Archeologické rozhledy, 74: 299-345.

In line with this is an increase in the importance of individual stallion lines, which is recognisable by the reduction in y-chromosome diversity. This was further increased after the Renaissance to the current extreme restriction. The genetic diversity of modern horse breeding has decreased by an average of around 16.4 % over the last 200 years, whereas it had remained fairly stable over the previous four thousand years of domestic horse history. [69]

Figure 5.1: Fully harmonised. Unusually high body control of horse and rider testify to an extremely close relationship between man and animal, which was made possible by the livestock genesis of the horse (riding demonstration by Ukrainian Cossacks).

---

[69] Molecular genetic findings according to:
Fages, A., Hanghoj, K., Khan , N., Gaunitz, C., Seguin-Orlando, A., Leonardi, M., McCrory Constantz, C., Gamba, C., Al-Rasheid, K. A. S., Albizuri, S., Alfarhan, A. H., et. al. & Orlando, L. (2019): Tracking Five Millennia of Horse Management with Extensive Ancient Genome Time Series. Cell, 177: 1419-1435.
Librado, P., Khan, N., Fages, A., Kusliy, M. A., Suchan, T. et al., & Orlando, L. (2021): The origins and spread of domestic horses from the Western Eurasian steppes. Nature, 598: 634-640. https://doi.org/10.1038/s41586-021-04018-9

5.2 Diversity

During the time that has passed since livestock genesis of the horse around 4,000 years ago, it was mainly used for labour. From the beginning, it had to fulfil two different basic functions, as a draught animal and as a pack animal. It was harnessed to pull carts, sledges and, for example, tree trunks, and it was used to carry a rider and other loads. In the course of breeding structural adaptation to these increasingly differentiated main functions, an enormous variety of breeds developed worldwide with huge differences in body size, conformation and temperament.

Figure 5.2: Akhal-teke, the epitome of high-legged, "fine-boned", "noble" horses of the oriental type, the fast riding horse.

Figure 5.3: Heavy cold-blooded horses in a team in front of a covered wagon, example of stocky, strong draught horses.

Figure 5.4. Dwarf ponies (Hellabrunn zoo, Munich) represent the lower extreme of horse size at just over half a metre in height.

5.3 Pseudo-wild horses

In view of the large number of horse breeds today and in the context of the central question of this book about the true nature of the Exmoor pony, it is necessary to consider the diversity of populations of wild living horses commonly referred to as wild horses. Probably the best known example are the mustangs of North America, a continent in which there were no real wild horses after the Ice Age. They are descendants of horses that escaped beginning in the 16th century during the Spanish conquest starting in Mexico and epitomise the diversity of Spanish horses half a millennium ago. Today's classic equestrian cultures of the Plains Indians only developed with the existence of these now wild living herds. They demonstrate how quickly the availability of horses suitable by their livestock character could influence and change human cultures not only in the Bronze Age, but also in modern times.

A similar example are the horses living wild in the Namibian Namib today. They only developed a century ago when the then German colony of South West Africa was abandoned as a result of the First World War from military horses left behind. They are therefore just as little true wild horses as the mustangs. In order to describe such populations with a catchy word, the term *pseudo-wild horses* is proposed here. This term is used in this book to describe horses that have been living in the wild in certain areas under the false flag as "wild horses" for a known or, in some cases, unknown period of time, but which have been proven or are at least highly likely to have originated from domestic horses that went feral at an early stage.

Inset 5.1: Wild animals and feral animals
When animals kept in captivity, being game or livestock, escape from constant supervision or are even deliberately released from it, they adapt as well as possible to the environment in which they have ended up. They run wild. If necessary, they build up their own populations in the local landscapes. If, in the course of such recolonisation, they encounter closely related forms to which there are no post-mating isolation mechanisms and pre-mating isolation mechanisms are only partially effective, this may result in the formation of hybrid zones or

even extensive hybrid populations. In the case of feral livestock, the nature of the livestock can be displaced by the nature of the game in this way. If there is no such hybridisation, the nature of the livestock remains intact even after feralisation. Livestock does not become game, even if it lives in the wild. Thus feral domestic horses do not become wild horses. Even if the term wild horses is commonly used for them, it is still incorrect and misleads us about the true nature of the animals. The only correct term for such horse populations is *pseudo-wild horses*. Wild horses reintroduced from captive breeding, on the other hand still remain genuine wild horses if they have not been crossbred with domestic horses during their time in captivity.

Figure 5.5: So-called wild horses that are not wild horses – horses grazing freely all year round in the Spanish Coto de Doñana National Park, surrounded by breeders. They are simple examples of pseudo-wild horses. The original Marismeño breed has become rather unremarkable Andalusian horses through further cross-breeding.

The most debatable pseudo-wild horses are small populations of small horses living in the wild in various places on the Iberian peninsula and in the northern Mediterranean region, some of which appear to be quite uniform and original. These include, among others, the Garranos in the north and the Sorraias in the south of Portugal. The most interesting of these two are the Sorraia horses, which represent type III in the hippological horse typology.

Figure 5.6: The two yellow dun and grey dun colour phases of the Portuguese Sorraia horses correspond to the wild colours originally found in European steppe wild horses (former breeding herd of Michael Schäfer).

After 1920, a conservation line was established from partly free living country horses from the Sorraia region with some particularly original-looking specimens, in which selection was started towards purity of type.[70] These horses are characterised, among other things, by their typical wild-colouring with yellow and grey duns, they have an eel line and in some cases a shoulder cross and leg stripes. Standardised in this way, this breed has even recently been suspected[71] to

---

[70] Nissen, J. (1999): Enzyklopädie der Pferderassen: Europa, Vol, 3. Stuttgart (Franckh-Kosmos).
Schäfer, M. (2000): Handbuch Pferdebeurteilung. Stuttgart (Franckh-Kosmos).
[71] Willmann, R. (2023): Arbeitsgruppe Ethologie, Ökologie und Evolution des Sorraia-Pferdes. https://www.uni-goettingen.de/de/sorraia-pferde/164770.html

represent surviving wild horses. Molecular genetics showed, however, that they belong to the oriental group of the domestic horse. Comparative behavioural studies (Chapter 6.2) also point to the domestic horse nature of the Sorraias, so that they must be seen as pseudo-wild horses that are essentially characterised by early domestic horse imports.

In order to be stylised as wild horses, two factors generally play a role for small populations of more or less free-living pseudo-wild horses. On the one hand, it is an origin that has been lost in the mists of history for many centuries, even though it is well known that foreign stallions of other breeds have been crossbred time and again. On the other hand, it is usually a predominance of the mouse-grey (gray dun) wild colour. A good example of the combination of both factors in Germany is the Westphalian Dülmener breed in the Merfelder moor. The repeated introduction of stallions of various breeds over the course of time,[72] even if once it may have been a relict wild horse population, resulted in an unmistakable grading-up to a domestic horse. It is therefore not at all surprising that, in terms of molecular genetics, the Dülmener can be categorised alongside the Sorraias in the oriental domestic horse group.

The breeding history of the Dülmener pseudo-wild horses was similar to that of some of the other British mountain and moorland populations in contrast to the Exmoor ponies. These include the Dartmoor and New Forest ponies (Figures 2.2, 2.3), without being categorised as wild horses. They too have repeatedly mixed up with other breeds, "improved" by breeding and thus increasingly stripped of their original character. [73]

The Polish Konik ("little horse") is a country horse that has never been wild but possibly still had a very small influx of wild horses at the beginning of the 19th century. Parallel to the creation of today's Sorraia horse in Portugal, a breeding stock of small horses was selected in Poland from the local country stock with the intention of breed-back the so-called forest tarpan (Chapter 4.4). This Konik breed

---

[72] Nissen, J. (1997): Enzyklopädie der Pferderassen: Europa, Vol, 1. Stuttgart (Franckh-Kosmos).

[73] Nissen, J. (1999): Enzyklopädie der Pferderassen: Europa, Vol. 3. Stuttgart (Franckh-Kosmos).

was ultimately misleadingly propagated under the name tarpan and thus became a pseudo-wild horse.[74]

Figure 5.7: Mouse-gray wild-coloured Konik, pseudotarpan, at the Polish Central Stud Farm of the Popielno Zoological Research Centre. Originally a small country horse selected for similarity to the forest tarpan, which in reality is not well known, and for a certain purity of type.

A second attempt to breed-back a tarpan, independent of the Konik, was carried out in the zoological gardens of Berlin and Munich. It was started by crossing Icelandic and Gotland mares with a Przewalski's stallion and then selected for the desired combination of traits.[75] In honour of the breeders, the horses created in this way are correctly called *Heck horses* in this book, instead of, erroneously, tarpans.

---

[74] Groves, C. P. (1994): Morphology, Habitat, and Taxonomy. In Boyd, L. & Houpt, K. A. (Eds.): Przewalski's Horse – The History and Biology of an Endangered Species: 39-59. New York (State University of New York Press).
Schäfer, M. (2000): Handbuch Pferdebeurteilung. Stuttgart (Franckh-Kosmos).
[75] Heck, H. (no year, before 1965): Rückzüchtung ausgestorbener Tiere. Sonderdruck

These are also pseudo-wild horses. With the intention of familiarising visitors with the former appearance of tarpans, some zoos exhibit Polish Koniks, others with the same intention Heck horses, and still others mixed herds of both breeds. Neither one nor the other is a tarpan. If the labeling of the herd nevertheless refers to tarpans, without an unambiguous additional explanation, this would be fraudulent labeling.

Figure 5.8: Heck horse, so-called bred-back tarpan, actually a pseudotarpan / pseudo-wild horse (Hellabrunn zoo, Munich).

For the Exmoor pony, the question remains as to whether it differs in character from pseudo-wild horses in the direction of true wild horses. Particular attention should be paid to breeds of the north-west European group to which it geographically is near.

Figure 5.9: Do Exmoor ponies have the livestock-like nature of pseudo-wild horses? (Photo Wolfgang Frey in Exmoor National Park)

6 Behavioural diagnostics

6.1 Jntroduction

As described in Chapter 3.1, the difference between game and livestock lies in their different natures. Livestock is docile compared to its wild ancestors, its behaviour appears subdued, characterised by a lesser significance of environmental influences.[76] Livestock or game characteristics can therefore only be identified by comparative-quantitative behavioural studies.

Based on the experience of the only project to date on new livestock breeding in fast motion,[77] a set of tools was compiled for another such project,[78] some of which can also be used for livestock or game diagnosis. It relates to behavioural changes that are fundamentally characteristic of livestock genesis. On the one hand, behavioural attenuation concerns individual reactions in different clearly defined situations. On the other hand, it concerns social interactions. For new livestock genesis, the reference background is the behaviour of the wild species, for a livestock-game diagnosis either the game or the livestock behaviour, possibly both reference bases.

This set of instruments for estimating breeding values in new livestock genesis contains the following tools: confinement acceptance, response to freeing, frightfulness, evasive distance, flight, individual distance when resting, individual distance when feeding, distance dispersion, vigilance, coordination. The first part of these instruments is usable for breeding selection under standardised experimental conditions, which are not available for the domestic horse – wild horse diagnosis of the Exmoor pony. The second group of instruments is free of such experimental approaches. It can therefore be used for behavioural characterisation in game – livestock comparisons.

---

[76] Hemmer, H. (1990): Domestication – The decline of environmental appreciation. Cambridge (Cambridge University Press)
[77] Hemmer, H. (2023): Neumühle-Riswicker – Viehwerdung im Zeitraffer. Norderstedt (BoD).
[78] Hemmer, H. & Carrasco, A. (2023): Erschaffung vielversprechenden Viehs für Lateinamerika. Norderstedt (BoD).

6.2 Individual distance

Close cohesion within a group of individuals of social mammals means for wild animals better protection for the group as a whole, whereas it is much less significant for livestock in a husbandry environment created and dominated by humans. Such cohesion can be quantified in the form of average distances between individuals, whereby the different basic activities of resting and feeding must be distinguished. A high or low individual variability of distances is also of interest. The combination of high social tolerance and low striving for social proximity, which characterises livestock, increases the arbitrariness and thus the variability of individual distances.

Differences between diverse horse breeds were described in the course of behavioural studies on Sorraia horses. "The wide individual distance between individual Sorraias ... stand in stark contrast to that of the Fjord horses and especially the Exmoor ponies which rarely keep a greater distance when grazing and whose suckling foals also remain in much closer contact with their mothers. The warm-blood and standard-bred mares and their foals behave quite differently, but are somewhat more similar to ponies." [79] From this comparative approach, a particularly close cohesion of the Exmoor ponies can be noted. The next smallest individual distances are found in Fjord horses, a breed of the north-western European group. High-bred horses, horses of the Oriental group, behave more randomly.

Breeders of Icelandic horses, like Fjord horses of the north-western European group of breeds, speak of their animals "sticking together" when it comes to their social behaviour.[80] However, if one compares them with Exmoor ponies, their cohesion is nevertheless rather loose (Figures 6.1-6.4). even if it is closer overall than in large horses of the Oriental group with increased distance dispersion. The wide individual distance of the Sorraia horses can be related to their higher aggressiveness.[80]

---

[79] Schäfer, M. (1986): Beobachtungen zum Verhalten des südiberischen Primitivpferdes (Sorraiapferd). PhD thesis, Veterinary Department, Ludwig Maximilians-University Munich.
[80] Ebhardt, personal communication

A study using rough measurements of Exmoor ponies, B-line Przewalski's horses (herd of B-line mares with a A-line stallion) and pseudotarpans (Heck horse mares with a Konik stallion) at Sababurg zoo confirms the very small individual distance between Exmoor ponies.[81] They were usually only 1-2 metres apart during periods of rest, and 10-15 metres apart during periods of activity. The Heck horses maintained very short distances, similar to the Exmoor ponies. In contrast, the B-line Przewalski's mares maintained double the distance to each other when grazing, usually more than 30 metres apart. This is in contrast to A-line Przewalski's horses, for which photo documentation shows a close community like in Exmoor ponies (Figures 6.5-6.6).

Figures 6.1-6.4: Individual distances in Exmoor ponies (left column; photos Wolfgang Frey in Exmoor National Park) present themselves much shorter than in Icelandic horses (right column on Icelandic pastures), although the latter are known for "sticking" to each other.

---

[81] Riediger, B. (1995): Untersuchungen zur Domestikation beim Exmoorpony. Diploma thesis, FB Biology, Johannes Gutenberg-University Mainz.

Figures 6.5-6.6: Very close grouping appears to be typical for A-line Przewalski's horses (top: Hellabrunn zoo Munich, left: Front Royal, Virginia).

Figure 6.7: Exmoor ponies show comparable behaviour to A-line Przewalski's horses in terms of their very short individual distance. (Photo Wolfgang Frey in Exmoor National Park)

Distances maintained between horses during the normal course of the day can be interpreted as an expression of various degrees of social bonding. Livestock usually gathers less closely together on average than game, a sign of subdued behaviour. On the one hand, they can be in an extremely close community, but on the other hand they can also be in a wide-open community, an expression of increased arbitrariness associated with a dampening of behaviour. The comparative observations on warm-blood horses are an example of this. In combination with increased aggressiveness, Sorraia horses are characterised by greater individual distances. Exmoor ponies, similar to Dzungarian wild horses (Przewalski's horses of the A-line) close together. Heck horses, with complex ancestry joining domestic as well as wild horses, are similar to Exmoor ponies.

6.3 Coordination

The striving for social bonding is expressed not only by individual distances, but also by social behavioural coordination and synchronisation within a group. The more dependent an animal's own behaviour is on the behaviour of other group members, the higher its desire to bond is to be rated. Attachment to the group can be highly necessary for the survival of game, whereas it is of much less importance for livestock in this respect. Determining the extent of social synchronisation is therefore another important tool for diagnosis on the game-livestock track.

The aforementioned comparative study of Exmoor ponies, B-line Przewalski's horses and Heck horses provides a general overview of the conformity of the Exmoor ponies and Heck horses when grazing and moving to drinking and resting places, whereas in the Przewalski's herd each horse largely acted on its own. Another study is based on the simultaneous recording of the behavioural categories of feeding, lying, rolling, standing, walking, trotting and galloping of all the mares observed together using minute logs. It allows a highly differentiated view of the behavioural coordination of horses of the northwest European group, which is characterised by smaller individual distances.

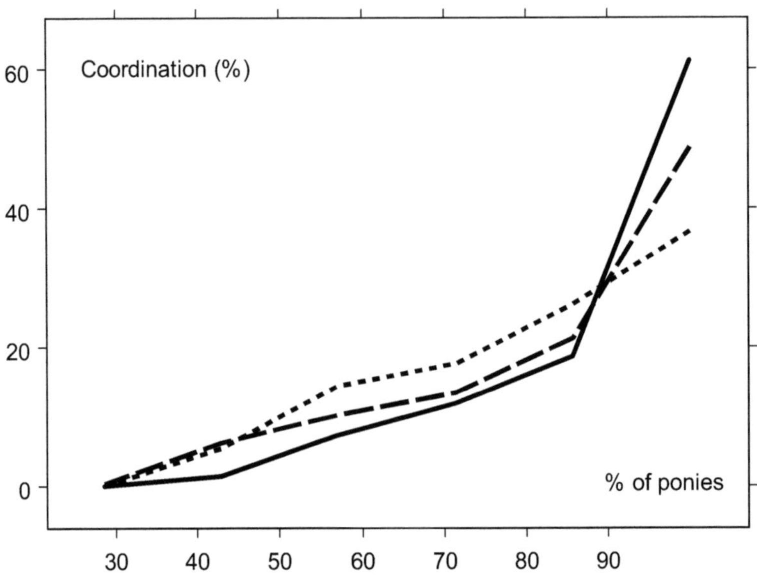

Figure 6.8: Differently close behavioural coordination in Exmoor ponies (unbroken line), Dartmoor ponies (long-dashed line) and Icelandic horses (short-dashed line). The figure shows the frequency with which a certain proportion (= % of ponies) of a herd of seven or eight mares behave in the same way (= coordination) in the same minute. 100 % of the ponies means that all individuals show the same behaviour in so many % of the minutes of observation. This is in the case for Exmoor ponies more than 60% of the time, for Dartmoor ponies only less than 50 % and for Icelandic horses even less than 40 %.

In this study,[82] groups of Exmoor ponies, Dartmoor ponies and Icelandic horses were each compared with summer and autumn observations that had sufficiently similar herd sizes and structures. The graphical illustration of behavioural coordination (Figure 6.8) is based on the evaluation of a total of 6462 differentially recorded observation minutes for Exmoor ponies, 8310 minutes for Dartmoor ponies and

---

[82] Siebert, K. (2003): Studien zur Frage des Wildpferd-Hauspferd-Charakters des Exmoor-ponys: Verhaltensstudien an verschiedenen Ponyrassen. Diploma thesis, FB Biology, Johannes Gutenberg-University Mainz.

7906 minutes for Icelandic horses. Despite similar curves for Exmoor and Dartmoor ponies, this very extensive data basis allows a clear distinction to be made between the two, with the Exmoor ponies proving to be the horses with the highest behavioural coordination. Figures 6.1-6.4 may help to understand the graphical presentation of herd synchronisation. Figure 6.1 shows 100 % coordination (walking), 6.2 and 6.3 each show 75 % coordination (3x standing, 1x grazing), 6.4 finally 40 % coordination (2x walking, 1x standing, 1x grazing, 1x lying down).

Exmoor ponies therefore behave like wild animals, comparable to the Dzungarian wild horse, which has remained unchanged through domestic horse crossbreeding, both in terms of their close grouping and their high level of behavioural synchronization. With their social behaviour, they have the nature of wild horses.

6.4 Attentiveness

Different levels of attention to their environment are expressed in the frequency with which individual animals check for threats. In the comparative study on Exmoor ponies, Heck horses and B-line Przewalski's horses the basic incidence of scanning the surroundings was recorded individually. All nine Exmoor pony mares scanned during the observation phases, but only one of eight Heck mares and none of the Przewalski's mares.[83] For these three herds, different opportunities for direct contact with the public (begging for food) considerably limit the significance of this observation, as only the Exmoor ponies were shielded in this respect. However, different reactions to sporadic appearance of other horses close to the enclosure boundary as potential disturbance stimuli that are identical for all groups give it a certain relevance. The Exmoor ponies backed-up briefly in such a situation, while the Heck horses did not react. This triggered strong territorial behaviour in the A-line Przewalski's stallion, while the B-line mares did not show any increased attention. Thus, with reservations, the scanning also underlines the findings of game-type behaviour of the Exmoor pony in comparison to other

---

[83] Riediger, B. (1995): Untersuchungen zur Domestikation beim Exmoorpony. Diploma thesis, FB Biology, Johannes Gutenberg-University Mainz.

horses. According to this criterion, the B-line Przewalski's horses observed in this study can no longer be regarded as wild horses.

6.5 Movement intensity

Dampening the behaviour of livestock compared to their respective wild ancestors usually involves a reduction in spontaneous motor activity.[84] However, this is not generally to be expected for the horse in particular, as great agility is of special interest to breeders for its use, at least in the strongly oriental-influenced riding horse breeds. A quantitative comparative study on Exmoor ponies, A-line Przewalski's horses and Trakehners, as an example of versatile warm-blood riding horses, aimed to provide initial answers to this complex of questions.[85]

The same behavioural categories were recorded as for the Exmoor, Dartmoor and Icelandic pony study (6.3), also in the minute protocol. There is a problem of comparability here, as the Exmoor ponies and one of the two groups of Przewalski's horses compared had free grazing all day, while the second Przewalski's group and the Trakehners were stabled at night. With regard to spontaneous locomotive activity, the difference in the number of minutes with locomotion, in many cases alongside other forms of activity, was statistically highly significant. The Przewalski's horses, which are comparable to the Exmoor ponies in terms of enclosure structure and management, had an average of 33 % minutes with locomotion during the daytime hours, while the Exmoor ponies had only 23 ½ %. As the Przewalski's horses stabled at night also showed 33 % minutes with locomotion, a cautious comparison with the Trakehners, which were also stabled at night, nevertheless seems possible. They had a locomotion frequency of 38 %, i.e. by no means less than the Dzungarian wild horses. Even if this result should be viewed with caution from a methodological point of view, it does indicate that the expectation that domestic horses, in contrast to other livestock, are particularly keen to move, is justified.

---

[84] Hemmer, H. (1990): Domestication – The decline of environmental appreciation. Cambridge (Cambridge University Press)

[85] Jendrosch, M.(1997): Studien zur Spontanaktivität von Pferden im Hinblick auf Haltung und Domestikation. Diploma thesis, FB Biology, Johannes Gutenberg-University Mainz.

Another aspect of this study again points to the wild animal side for the Exmoor pony. The generally lower intensity of movement to be assumed for livestock than for game in the context of their behavioural attenuation is also expressed here. Since spontaneous locomotion at trot and gallop together account for less than 10 % of movement at walk in all the horses compared, the movement performance is only considered comparatively for the walk. Trakehners took an average of 10 ½ steps per minute of locomotion, Przewalski's horses took 13 ½ and Exmoor ponies 14 ½. Here the Exmoor pony is clearly on the wild horse side.

Figure 6.11: With their social behaviour. attentiveness and intensity of movement, Exmoor ponies do not demonstrate the behaviour of domestic horses, but that of wild horses, with close cohesion, high coordination, great attentiveness and intense movement. (Photo Wolfgang Frey in Exmoor National Park)

7 The jigsaw puzzle

The basic finding that the Exmoor pony has the nature of a wild horse and can therefore be regarded as such is not in itself sufficient to establish its indispensable connection to a specific population or population group of *Equus ferus* that was believed to be extinct. In order to complete the full picture, there must be no contradictions with the results of molecular genetics, the brain size must be within the known range and it must be possible to prove complete evolutionary coherence with palaeontological knowledge. The following subchapters will shed light on this.

7.1 Molecular genetic evidence

In the meantime, an enormous number of genomes from both modern horse populations and DNA samples from both fossil and sub-fossil bone material have been sequenced, the analysis of which has provided decisive insights into the relationships between wild and domestic horses (Chapters 4 and 5). However, such studies are still lacking for British mountain and moorland ponies, including the Exmoor pony. In terms of molecular genetics, it is therefore necessary to fall back on the one-sided status quo of analyzing mitochondrial DNA because it only represents dam lines. The most common mitochondrial lines (haplotypes) in Exmoor ponies are already found in Late Pleistocene and early Holocene in western continental Europe, which lived before livestock genesis. None of the other haplotypes detected in Exmoor ponies can be interpreted as later introgression of domestic horses.[86]

---

[86] Jansen, T.,Forster, P., Levine, M. A., Oelke, H., Hurles, M.,Renfrew, C., Weber, J., & Olek, K. (2002): Mitochondrial DNA and the origins oft he domestic horse. PNAS, 99, 16: 10905-10910.
Cieslak M, Pruvost M, Benecke N, Hofreiter M, Morales A, Reissmann M, et al. (2010): Origin and History of Mitochondrial DNA Lineages in Domestic Horses. PLoS ONE 5(12): e15311. https://doi.org/10.1371/journal.pone.0015311
Hovens, J. P. M. & Rijkers, A. J. M. (2013) : On the origins of the Exmoor pony: did the wild horse survive in Britain? Lutra, 56, 2: 129-136.

7.2 Brain size

Today, there is a lack of data to directly determine the relative brain size of the Exmoor pony from the calculation of body and brain masses. Values determined indirectly from the measurement of the size and brain capacity of skulls show a scattering in the middle ranges of domestic horses, below that of Przewalski's horses.[87]

Comparative data from European wild horses are extremely sparse, but when analysed together they are nevertheless meaningful. A Late Pleistocene English horse from the Greater London area (Lea)[88] fits into the central range of domestic horse skulls. It proves that in late glacial western European populations there were at least individuals with a brain size no larger than that of the average domestic horse. This finding is supported by horse skulls, unfortunately not C^{14} dated, which were recovered from gravel quarries on the Upper Rhine in Rhineland-Palatinate and catalogued as "Rheindiluvium".[89] One of the four skulls that allowed relevant measurements should be of late-glacial age, as it is in a comparable state of preservation to an Ice Age bison (*Bison priscus*) from the same location. The others appear rather to date from post-glacial millennia. All skulls have relative brain sizes in the middle of the domestic horse range.

The skull of one of the last Ukrainian tarpans provides a third confirmation of the finding of indistinguishable brain sizes of domestic and European wild horses. It is also in the range of domestic horses, which was initially interpreted as an indication of possible hybridisation.[90] To summarise, Exmoor ponies have somewhat smaller brains than Przewalski's horses, but there is no evidence of a reduction in their brain size compared to at least Western European wild horses. This finding is inevitably linked to the earlier hypothesis that the horse is possibly the only species whose livestock genesis was not accompanied by a reduction of brain size. However, it also con-

---

[87] Riediger, B. (1995): Untersuchungen zur Domestikation beim Exmoorpony. Unpublished diploma thesis, Institute of Zoology, FB Biology, Johannes Gutenberg-University Mainz.

[88] Lea Drift, British Museum of Natural History, Palaeontological Collection, BMNH 16170.

[89] Museum of Natural History, Mainz, Inventory numbers 1952/189, 1953/109,1955/102, 1968/320

[90] Gorgas, M. (1966): Betrachtung zur Hirnschädelkapazität zentralasiatischer Wildsäugetiere und ihrer Hausformen. Zoologischer Anzeiger, 176: 227-235.

firms again that in the presence of populations with different brain sizes, livestock genesis did not originate from large-brained forms.[91]

7.3 Mane

The shape of the mane was long regarded as a reliable criterion for distinguishing between wild and domestic horses. All wild horses were said to have upright manes, while all domestic horses had hanging manes that fell to the side. At the latest since the discovery 30 years ago of an approximately 30,000 year old horse mummy with a long mane in the North American permafrost of the Yukon region, this view had already become out-dated.[92]

The extraordinarily high number of horse depictions in European Ice Age art refers to a large extent to upright manes, but also includes, albeit sparse, references to hanging manes. Falling manes are repeatedly and unequivocally documented in the cave of La Pasiega (Cantabrian mountains, northern Spain). The hunted game depicted here over a long period of time (Aurignacian to Magdalenian) is dominated by horses and red deer. Bison, aurochs and ibex are represented to a comparatively small extent. [93] The depictions of La Pasiega prove that the characteristic hanging mane existed in late-glacial horse populations of the Franco-Cantabrian mountain landscapes.

7.4 Diagnostic structures

The shape of the metapodials and the implantation and size of the mandibular premolars and molars shown in X-ray images link Exmoor

---

[91] Hemmer, H. (1990): Domestication – The decline of environmental appreciation. Cambridge (Cambridge University Press).
Groves, C. P. (1994): Morphology, Habitat, and Taxonomy. In Boyd, L. & Houpt, K. A. (Eds.): Przewalski's Horse – The History and Biology of an Endangered Species: 39-59. New York (State University of New York Press).
[92] Zazula, G. & Froese, D. (2011): Ice Age Klondike – Fossil treasures from the frozen ground. Government of Yukon.
[93] Evaluation of the comprehensive collection of illustrations up to the early 1960s in Müller-Karpe, H. (1966) Handbuch der Vorgeschichte, Vol. 1. München (Beck).
See also Fig. 54 /A) in Eisenmannn, V. (2022): Old World Fossil *Equus* (Perissodactyla, Mammalia), Extant Wild Relatives, and Incertae Sedis Forms. Quaternary, *5*, 38. https://doi.org/10.3390/ quat5030038

ponies to Pleistocene English horses. With regard to these two skeletal features used for diagnosis, they are contrasted with horses from Iron Age English archaeological sites, which are labeled as Celtic ponies. These small horses of the Celtic culture of England have nothing in common with the Exmoor pony.[94]

The width of bones of the foot skeleton is not only of diagnostic importance in the hippological horse typology, but also in equine palaeontology for mainly Western and Central European horses as opposed to Eastern European steppe horses of the Late Pleistocene. These two different approaches, which primarily work independently from each other, ultimately point to a consistent result. The *pony types* refer to Western and Central European horses, the *horse types* are based on horses from the steppes of the east. The conspicuous width of individual elements of the foot skeleton is reflected in different names given by the two approaches: broad-hinged epiphyses of the metacarpals, *Equus ferus latipes* = broad-footed wild horse, broad-hoofed horse.

An index (robustness index) frequently used in palaeontology consisting of the greatest length and the smallest shaft width (diaphyseal width) of the metacarpus can lead to misinterpretations with regard to this characteristic if pieces of very different sizes are compared with each other. Allometric calculations and width-length diagrams provide more reliable information in such cases. This form of comparative visualisation is chosen here. The few metacarpals available from adult Exmoor ponies (one stallion, one mare, each of medium size) are positioned near the minimum dimensions of the latest glacial *Equus ferus arcelini* and prove to originate from broad-footed horses (Figure 7.1).

The measurement of the only existing skeleton of a Ukrainian Tarpan (*Equus ferus gmelini*) is located next to the Exmoor pony values. Individual finds of Celtic-Germanic horses and Late Pleistocene Portuguese *Equus ferus antunesi* are also found near the Exmoor ponies. However, these two groups as a whole are shifted to narrower metacarpals and thus are no broad-footed horses. The domestic horses

---

[94] Speed, J. G. & Speed, M. G. (Eds./1977): The Exmoor-Pony. Chippenham (Countrywide Livestock Ltd).

of the north-west European group of breeds have on average even less broad metacarpals than the Portuguese wild horses. This diagram emphasizes the previously not metrically proven finding that the Exmoor pony is assigned to Pleistocene horses and does not belong to Celtic ponies.[95]

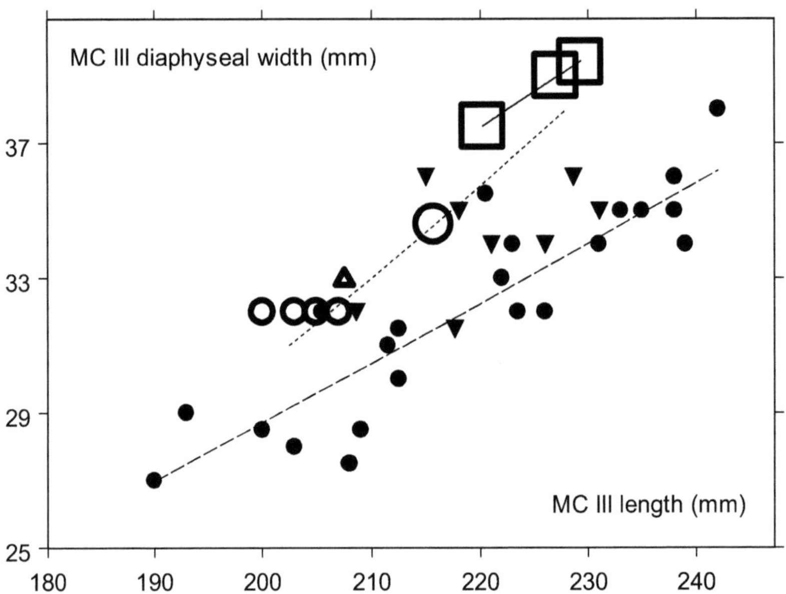

Figure 7.1: Diagram showing the dependence of the smallest diaphyseal width on the greatest length of the metacarpus(MC III) of wild and domestic horses. Open symbols: broad-footed horses; large open squares: mean values of large series of three populations of Equus ferus germanicus (including Equus ferus gallicus) from French sites of different chronological age (from right to left corresponds early to late) with corresponding allometric line[96]; large open circle: mean of latest glacial Equus ferus arcelini, with dotted line connecting mini-

---

[95] Speed, J. G. & Speed, M. G. (Eds./1977): The Exmoor-Pony. Chippenham (Countrywide Livestock Ltd).

Ebhardt, H. (1962): Ponies und Pferde im Röntgenbild nebst einigen stammesgeschichtlichen Bemerkungen dazu. Säugetierkundliche Mitteilungen,10: 145-168.

[96] Guadelli, J.-L. (1991): Les chevaux de Solutré (Saône et Loire, France). Cahiers du Quaternaire, 16 : 261-336.

mum and maximum valus[97]; small open circles: Exmoor ponies[97]; open triangle with tip pointing upwards: Ukrainian tarpan (Equus ferus gmelini)[98]; closed triangles with tip pointing downwards: Equus ferus antunesi[99]; closed circles: Gallo-Roman and Germanic horses with associated allometric curve (interrupted line) [100].

The relative size of the upper tooth row allows different lines of development to be distinguished. With comparable skull proportions, an increase of the size of the cheek teeth is at the expense of their distance (diastema) from the outer incisors, i.e. from the front teeth. The larger the former are, the smaller the ratio of the length of the diastema to the length of the cheek tooth row. For Exmoor ponies, this ratio (length of the diastema in % of the length of the tooth row) is 51 % on average, with a variation of a small series of 45-57 % (n = 6).[101] Mares with relatively smaller teeth range above the mean, stallions with relatively larger teeth are below.

Even smaller values are found in Przewalski's horses. They have comparatively huge teeth, which is functionally understandable with their sand-loaded and thus more abrasive natural food supply. The holotype skull of the older broad-footed horse *Equus ferus germanicus* has relatively smaller teeth than the Exmoor ponies, while a chronologically younger skull assigned to the same subspecies lies in their range, which also applies to the Ukrainian tarpan. [102]

---

[97] Biodiversity Museum Göttingen; measured by Gisa Heinemann

[98] Bibikova (1970) cited by Anthony, D.W. (1991): The Domestication oft he Horse. In Meadow, R. H. & Uerpmann, . P. (Eds.): Equids in the Ancient World, Volume 2: 250-277. Wiesbaden (Reichert).

[99] Cardoso, J. K. & Eisenmann, V. (1989): *Equus caballus antunesi*, nouvelle sous-espèce Quaternaire du Portugal. Palaeovertebrata, 19, 2: 47-72.

[100] Hemmer, H. & Jaeger, R. (1969): Über ein Pferdeskelett aus dem fränkischen Gräberfeld bei Eltville (Rheingau)nebst Bemerkungen zur Abstammung der Hauspferde. Zeitschrift für Tierzüchtung und Züchtungsbiologie, 85, 3: 221-241.
Hemmer, H. & Jaeger, R. (1969): Pferde zur Römerzeit im Mainz. Mainzer Naturwissenschaftliches Archiv, 8: 347-362.
Nobis, G. (1973): Zur Frage römerzeitlicher Hauspferde in Zentraleuropa. Zeitschrift für Säugetierkunde, 38: 224-252.
Schaal, F. (1968): Tierknochenfunde aus der Siedlung „Am Hetelberg" bei Gielde/Niedersachsen. PhD thesis, Ludwig Maximilians-University Munich.

[101] Biodiversity Museum Göttingen; Scans: Gisa Heinemann.

[102] Values for these horses estimated from photos of

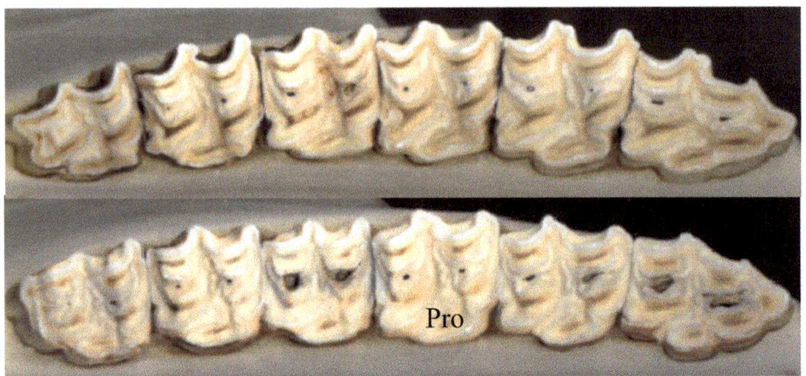

Figure 7.2: Variability of relative length and shape of the inner pillar (protocone) in the maxillary teeth of Exmoor ponies in occlusal view. Teeth from left to right: M^3, M^2, M^1, P^4. P^3, P^2; M = molar, P = premolar, Pro = protocone. Top stallion (ZMUG 23413), bottom mare (ZMUG 8274). The two rows of teeth brought to the same length also demonstrate the difference in size between the individual teeth of stallions and mares.

Table 7.1: Protocone proportions of the upper cheek teeth of last glacial horses[103] in comparison to Exmoor ponies (complete tooth rows) [104] and to Roman-period Central European domestic horses[105]; supplemented by a Ukrainian tarpan[106]. Splitting subspecies taxonomy with Equus ferus germanicus, E. ferus gallicus and E. ferus arcelini. Means and variabilities, for E. ferus antunesi and Exmoor ponies also standard deviations, in brackets number of teeth or tooth rows.

Eisenmannn, V. (2022): Old World Fossil *Equus* (Perissodactyla, Mammalia), Extant Wild Relatives, and Incertae Sedis Forms. Quaternary, 5, 38. https://doi.org/10.3390/ quat5030038

[103] Guadelli, J.-L. (1991): Les chevaux de Solutré (Saône et Loire, France). Cahiers du Quaternaire, 16 : 261-336.

Cardoso, J. K. & Eisenmann, V. (1989): *Equus caballus antunesi,* nouvelle sous-espèce Quaternaire du Portugal. Palaeovertebrata, 19, 2: 47-72.

[104] Dlmensions determined from greatly enlarged scans of the occlusal surfaces of the maxillary teeth of the skulls of Exmoor ponies from the Göttingen Biodiversity Museum. . Scans Gisa Heinemann.

[105] Measurements by Nobis, G. (1973): Zur Frage römerzeitlicher Hauspferde in Zentraleuropa. Zeitschrift für Säugetierkunde, 38: 224-252.

[106] Eisenmann, V. (1991): Les chevaux Quaternaires Européens (Mammalia, Perissodactyla). Taille, typologie, biostratigraphie et taxonomie. Geobios, 24, 6: 747-759.

Locality, age (C^{14}- dating), taxon assignment	$P^{3/4}$ protocone in % tooth length	$M^{1/2}$ protocone in % tooth length	$P^{3/4}$ protocone in % $M^{1/2}$ protocone
Combe-Grenal ≥ 35.000 *E. f. germanicus*	45,7 36,1-62,1 (71)	52,7 40,7-68,0 (70)	97,6
Camiac ~ 35.000 *E. f. gallicus*	45,5 40,3-55,3 (22)	53,6 45,4-60,8 (33)	95,0
Solutré ~ 23.500 *E. f. gallicus*	47,4 40,7-56,1 (20)	55,7 40,2-66,0 (22)	94,1
Solutré ~ 12.500 *E. f. arcelini*	49,4 42,9-57,1 (19)	54,0 43,1-64,6 (81)	103,2
Salemas, Fontainhas, João Ramos ~29.000- ~14.000 *E. f. antunesi*	47,3 ± 3,9 41,9-56,6 (11)	50,8 ± 3,5 44,8-57,7 (14)	104,7
Exmoor pony	47,9 ± 4.9 38-56 (28)	53,4 ± 3,7 46-59 (28)	102,1 97,0-105,7 (14)
Kherson region 19th century AD. *E. f. gmelini*	39,2 (2)	45,3 (2)	104,6
Krefeld-Gellep 1st century AD *E. caballus:* Celtic-Germanic group of breeds	41,7 30,6-50,3 (33)	48,5 36,6-.60,2 (34)	96,7

The upper dentition makes it extremely unlikely that the Exmoor pony is related to Roman period Central European domestic horses of the Celtic-Germanic breed group with a short protocone. It is placed with the palaeontological Eisenmann tooth type II on the one hand in the vicinity of the Portuguese *Equus ferus antunesi* and on the other hand very close to the late Magdalenian Solutré horses (Burgundy, France), the youngest end-Pleistocene Western European wild horse population (*Equus ferus arcelini*).

In the same period and geographically not far away, such a pony was perfectly portrayed by an Ice Age artist in the cave of Niaux in the French Pyrenees. What we see there is in fact nothing other than an enlarged Exmoor pony in its winter coat (Figure 7.3). Horse bones from the Mendip caves in the south-west of England have also been dated to this period.[107] Finds from this cave system played a major role in recognising similarities with Exmoor ponies.[108]

The diagnostic structures of both the metacarpus and the dentition once again confirm the earlier finding[109], that Exmoor ponies have nothing to do with small horses of the Iron Age Celtic-Germanic-Roman cultures. It is doubtful whether domestic horses came to the north-west European islands at all before the Iron Age.[110] In any case, the idea that the Exmoor ponies are feral Celtic domestic horse imports, which has been repeatedly expressed without evidence, should be relegated in the realm of myth.

7.5 European mountain pony – *Equus ferus arcelini*

The facts summarised here leave no doubt that Exmoor ponies are to be understood as Western European broad-footed horses that survived the Pleistocene and did not undergo any demonstrable change in character from game to livestock. Following livestock genesis of the horse, occasional introgression of domestic horse genes over the last four millennia could be expected. They appear to be far less problematic than in re-introduced Przewalski's horses of the B-line, which have been positively selected for domestic horse genes for half a century.

---

[107] Burleigh, R. (1986) : Radiocarbon dates for human and animal bones from Mendip caves. Proc. Univ. Bristol Spelaeol. Soc., 17, 3: 267-274.
Burleigh, R., Currant, A., Jacobi, E., & Jacobi, R. (1991): A Note on some British Late Pleistocene Remains of Horse (*Equus ferus*). In Meadow, R. H. & Uerpmann, . P. (Eds.): Equids in the Ancient World, Volume 2: 233-237. Wiesbaden (Reichert).
[108] Speed, J. G. & Etherington, M. G. (1952): An Aspect of the Evolution of British Horses. The British Veterinary Journal, 108, 5. Nachdruck in Speed, J. G. & Speed, M. G. (Eds./1977): The Exmoor-Pony, 14-25. Chippenham (Countrywide Livestock Ltd).
Ebhardt, H. (1962): Ponies und Pferde im Röntgenbild nebst einigen stammesgeschichtlichen Bemerkungen dazu. Säugetierkundliche Mitteilungen,10: 145-168.
[109] Speed, J. G. & Speed, M. G. (Eds./1977): The Exmoor-Pony. Chippenham (Countrywide Livestock Ltd).
[110] Bendrey, R., Thorpe, N., Outram, A., & van Wijngaarden-Bakker, L. H. (2013): The Origins of Domestic Horses in North-west Europe: new Direct Dates on the Horses of Newgrange, Ireland. Proceedings of the Prehistoric Society, 79:

> **Inset 4.1: Livestock genes in wild animals**
>
> An illustrative example from the game-livestock comparison of domestic animal crossbreeding in game populations in their natural habitat that does not lead to a fundamental loss of game character is provided particularly clearly by the North American bison. In all surviving and reintroduced populations of this undisputed wild cattle in North America today, there are probably no individuals left without a small proportion of domestic cattle genes.[111]

For the Exmoor pony, this raises the question of its phylogenetic position and its naming within the framework of zoological nomenclature. This question has always remained unanswered in publications with a wild horse reference (Chapter 2). The Exmoor pony has nothing to do with the Russian tarpan (*Equus ferus ferus*) and the Ukrainian tarpan (*Equus ferus gmelini*). The same applies to the Central Asian Dzungarian wild horse (*Equus ferus przewalskii*). The Late Pleistocene western Iberian wild horse (*Equus ferus antunesi*), which is similar to the Exmoor pony in terms of its dentition, but is not a broad-footed horse, can also be ruled out. The Polish, possibly also Central European forest tarpan (*Equus ferus silvestris*), which only became extinct at the beginning of the 19th century, could come into consideration due to its apparently high ecological plasticity in view of its former forest habitat. Apart from its colour, hardly anything is known about it, but just its mouse-grey colouring with an eel line shows that this population did not correspond to the Exmoor pony.

Regarding the survival in Western Europe, we must therefore look at the population that was last widespread here at the end of the Ice Age. It does not matter whether it is documented in England or in France or in both countries. The English south was ice-free in the last glacial period and, thanks to the strong lowering of the sea level, was connected to the continent to form a continuous possible settlement. The broad-footed horses of Western Europe already became smaller during the Late Pleistocene. Their first chronopopulations (*Equus*

---

[111] Stroupe, S., Forgacs, D., Harris, A., Derr, J. N., & Davis, B. W. (2022): Genomic evaluation of hybridization in fistoric and modern North American Bison (*Bison bison*). Scientific Reports, 12:6397. https://doi.org/10.1038/s41598-022-09828-z

ferus germanicus) represented the maxillary cheek tooth type I. They were eventually replaced by even smaller ponies with type II teeth. Their end-glacial population of the Burgundy region (*Equus ferus arcelini*) corresponds perfectly to the Exmoor pony with the diagnostic structures of both their maxillary molars and premolars as well as their metapodials, which merely continued the reduction in body size that began in the late Ice Age. There is no reliable evidence for the coat colour of these horses. Coloured cave paintings are dated earlier, i.e. relate to the predecessor population. Although the Exmoor ponies are significantly smaller on average than the latest glacial French horses, their above-medium sized individuals scatter in the below-medium sized range of those Ice Age horses. Thus, the 75 % rule of subspecies differentiation based on this characteristic alone does not allow a taxonomic separation.

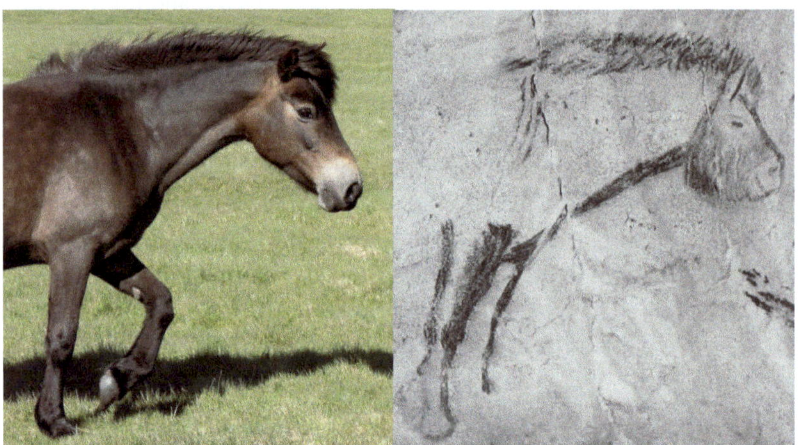

Figures 7.3-7.4: Exmoor pony (photo Wolfgang Frey in the Exmoor National Park) in comparison with a representative of the late Magdalenian population of the European mountain pony (Equus ferus arcelini) of the northern Pyrenees (Niaux cave)[112]: same proportions, mane with large forelock, hint of mealy muzzle and eye ring. The late Ice Age pony depicted has the beard of the winter coat and appears somewhat stockier than the Exmoor pony, which is easily understood from its increased body size.

---

[112] L'Art Préhistorique – peintures, gravures et sculptures rupestres (1951). Paris (Braun).

Figures 7.5-7.8: Late Magdalenian large mammal community in the northern Pyrenees, hunting game of the Ice Age artists of the Niaux cave: Exmoor pony, current relict of the late glacial European mountain pony – Equus ferus arcelini (photo Wolfgang Frey); Pyrenean ibex – Capra pyrenaica (photo Wolfgang Frey); North American prairie bison – Bison bison, as a current relative of the late glacial Eurasian steppe bison; red deer – Cervus elaphus.

The broad-footed horse population that ultimately survived on Exmoor and with it the entire original population of British mountain ponies[113] before the livestock genesis of the horse can therefore be taxonomically linked to the end-glacial wild horse subspecies *Equus ferus arcelini* as its further size-reduced, modern-day relict, which has been preserved for over twelve millennia. Based on the English naming of this population group, which today is only preserved with the Exmoor ponies, this Late Pleistocene to present-day subspecies may be better called *European mountain pony* with an ecological

___

[113] Baker, S. (2008): Exmoor-Ponies – Survival of the Fittest – A Natural History. Revised and updated edition. Chippenham (Exmoor Books).

reference instead of the morphologically very characteristic but less catchy name broad-footed horse.

According to the cave art of Niaux, the late Ice Age mountain ponies lived together with red deer, steppe bison and Pyrenean ibex in the Franco-Cantabrian south of their distribution. In principle, this ecological community is comparable to that of today's Exmoor. Here, too, the mountain pony population lives together with a population of red deer. The steppe bison, which became extinct at the end of the last Ice Age, as well as the ibex, have been ecologically replaced by cattle and sheep in the Exmoor landscape.

Figure 7.9: Exmoor pony – surviving relict of the European mountain pony – Equus ferus arcelini. (Photo Wolfgang Frey in the Exmoor National Park)

8 The livestock genesis of the horse

The fact that Exmoor ponies are not livestock but game, the last surviving European mountain ponies, makes knowledge on the handling of horses in human hands even more complex than it already is. The horse phenomenon seems to be only half understood. It is usually assumed that horses were originally animals of wide open landscapes. The attempt was made to understand all horses from this bias. European mountain ponies, however, are ecologically much more adaptable. They are inhabitants of hilly, mountainous and even rocky landscapes, inhabitants of both wooded and sparsely wooded habitats, unexpectedly adaptable animals. The thesis that progressive forestation must lead to the disappearance of wild horses[114] should be reconsidered in view of this broader ecological adaptation of European mountain ponies. Forest tarpans (Chapter 4.4) would then also have been unthinkable.

The inclusion of European mountain ponies in the ancestry of modern horse breeds suggests a modified view on the evolution of the horse. For over a century[115] it has been pointed out on the basis of different and progressively extended methods that domestic horses are of polyphyletic origin. With the knowledge of molecular genetics, polyphyletic means genetically fed from more than a single wild horse population.[116] However, this is anything but unique for livestock.[117] Wherever native wild horse populations have existed, they have left

---

[114] Warmuth, V. , Eriksson, A., Bower, M.A., Canon, J., Cothran, G., Distl, O., Glowatzki-Mullis, M.-L., Hunt, H., Luis, C., do Mar Oom, M., Tupac Yupanqui, I., Ząbek, T., & Manica, A. (2011): European Domestic Horses Originated in Two Holocene Refugia. PLOS ONE 6(5): https://doi.org/10.137/journal.pone.0018194
Sommer, R. S., Benecke, N., Löugas, L., Nelle, O., & Schmölcke, U. (2011): Holocene survival of the wild horse in Europe: a matter of open landscape? Journal of Quaternary Science, 26, 8: 805-812.

[115] Ewart, J. C. (1904): The multiple origin of horses and ponies. Nature 69, 1799: 590-596.

[116] Cieslak M, Pruvost M, Benecke N, Hofreiter M, Morales A, Reissmann M, et al. (2010): Origin and History of Mitochondrial DNA Lineages in Domestic Horses. PLoS ONE 5(12): e15311. https://doi.org/10.1371/journal.pone.0015311
Librado, P., Khan, N., Fages, A., Kusliy, M. A., Suchan, T., et al. & Orlando, L. (2021): The origins and spread of domestic horses from the Western Eurasian steppes. Nature, 598. https://doi.org/10.1038/s41586-021-04018-9

[117] Hemmer, H. & Carrasco, A. (2023): Erschaffung vielversprechenden Viehs für Lateinamerika. Norderstedt (BoD).

genetic traces after the introduction of domestic horses. In Europe, this has been proven by molecular genetics for both the Balkan region and the Iberian Peninsula.[118] In all these cases, it is not a question of primary livestock genesis.[119]

Nevertheless, the evolution of the horse as a livestock is indeed a unique story. The preadaptations found in European wild horse populations, the eastern steppe horses, the western mountain ponies and possibly also the Iberian steppe horses – in contrast to the larger-brained Dzungarian steppe horses – appear extraordinary. When caught young, such wild horses can behave like docile livestock towards their owners without having undergone livestock genesis. In this respect they appear comparable to working elephants.

After a phase that can be estimated to have lasted up to two millennia, during which horses were not only intensively used for hunting, but certainly also kept, the actual livestock genesis followed. According to the current state of knowledge, this was not a slow process of small steps, as with other livestock, but obviously took place surprisingly quickly. In contrast to other livestock, nothing needed to change in terms of brain size. Preadaptations in coat colour apparently played a leading role in the final stage of the livestock genesis process. They came to the fore within a few centuries. It can be assumed that changes in character associated with different coat colours were even more significant in the comparatively rapid livestock genesis of the horse than in the evolution of other livestock.

The extremely wide colour palette of the Icelandic horse population, which has remained free of imports for a millennium, provides an insight into the high variability of the Celtic-Germanic breed group at the end of the first millennium AD, long before modern times.

---

[118] Dzhebir, G., Yordanov, G.,, Yankova, I., Sirakova, D., Petrova, M., Neov, B., Hristov, P., Radoslavov, G., Hristova , L., & Spassov, N. (2018): Comparative genetic analysis of subfossil wild horses (from the Neolithic Age and Early Bronze Age) and present-day domestic horses from Bulgaria. Historia naturalis bulgarica, 25: 3–10.
Lira, J., Linderholm, A., Olaria, C., Brandström Durling, M., Gilbert, M. T. P., Ellegren, H., Willerslev, E., Lidén, K., Arsuaga, J. L., & Götherström, A. (2010): Ancient DNA reveals traces of Iberian Neolithic and Bronze Age lineages in modern Iberian horses. Molecular Ecology, 19: 64-78.
[119] For the terms primary and secondary domestication, see: Hemmer, H. (1990): Domestication – The decline of environmental appreciation. Cambridge (Cambridge University Press).

Figures 8.1-8.8: Small sample of the huge colour palette of horses on Icelandic pastures, including the wild colours of the Exmoor pony and the tarpan (below).

In addition to a large number of colour mutants in various combinations, all the wild colours that existed in Europe can still be found here.[120] They prove that not only mouse-grey steppe horses played a role in the genesis of this breed group, but also European mountain ponies of the Exmoor pony colour type.

The existence of the Exmoor pony as a relict of Western European wild horses provides an insight into the nature of native European mountain ponies at the time when the first domestic horses were brought westwards. Mountain pony mares captured as foals, or at least at a very young age, should have been fairly easy to control and should not have posed any problems for keeping them. It should therefore have been easy to build up large horse populations with rapidly increasing genetic proportions of imported horses within half a century by mating them with imported domestic stallions in the form of grading-up displacement breeding. Such an approach produces animals with mitochondrial DNA from the regional dams that largely correspond to the domestic horse type throughout the genome. This scenario makes it possible to understand the great uniformity of all domestic horses on the one hand and their great diversity expressed on the maternal side on the other. It can be assumed that the molecular-genetically speciality of the north-west European breed group of the domestic horse[121] developed in this way through the involvement of north-west European mountain ponies. Their contribution to the differentiation of cold-blooded breeds[122] should also be expected in view of specific characteristics.

---

[120] Magnussón, S.A. & Þorkelsson, F. (2001): Das Islandpferd und seine Farben. Reykjavík.

[121] Fages, A., Hanghoj, K., Khan , N., Gaunitz, C., Seguin-Orlando, A., Leonardi, M., McCrory Constantz, C., Gamba, C., Al-Rasheid, K. A. S., Albizuri, S., Alfarhan, A. H., et. al. & Orlando, L. (2019): Tracking Five Millennia of Horse Management with Extensive Ancient Genome Time Series. Cell, 177: 1419-1435.

[122] Hemmer, H. & Jaeger, R. (1969): Über ein Pferdeskelett aus dem fränkischen Gräberfeld bei Eltville (Rheingau) nebst Bemerkungen zur Abstammung der Hauspferde. Zeitschrift für Tierzüchtung und Züchtungsbiologie, 85, 3: 221-244.

Schäfer, M. (2000): Handbuch Pferdebeurteilung. Stuttgart (Franckh-Kosmos).

Inset 8.1: Neumühle-Riswicker deer – experimental archaeology model of the spread of early livestock

The livestock genesis of the Neumühle-Riswicker deer from the European fallow deer[123] not only demonstrate selection processes in this evolutionary event as an experiment in a time lapse of two orders of magnitude, but also illustrates what happened during the spread of early livestock across areas occupied by the ancestral species.

Even during the first phase of the project, visiting farmers at the Rhineland-Palatinate Teaching and Research Centre for Livestock Husbandry Neumühle expressed their interest in yearling males from the new breeding line. Two decades after the start, the success in the two institutions now involved (Neumühle and Haus Riswick Agricultural Centre) was obvious. As a result, the demand of young bucks of the new breed rose sharply. The farmers who used them for breeding improvement with their herds of European fallow deer and recognised the positive result in facilitating the handling of the animals in increasing their production through crossbreeding continued this procedure. Even those who did not strictly follow the rules of grading-up displacement breeding ensured a progressive increase in the proportion of Neumühle-Riswicker in their herd in this way. Within a very short time, this resulted in a continuously small number of purebred animals, but a large number of different mixes.

Molecular genetic findings on various livestock are consistent with this experimental archaeology model for the course of livestock genesis in connection with secondary domestication when genetically different wild populations are available.[124] This explanatory model can also be used to interpret molecular results on the domestication of horses.

---

[123] Hemmer, H. (2023): Neumühle-Riswicker – Viehwerdung im Zeitraffer. Norderstedt (BoD).

[124] Compilation for the livestock of the world in:
Hemmer, H. & Carrasco, A. (2023): Erschaffung vielversprechenden Viehs für Lateinamerika. Norderstedt (BoD).

Figure 8.9: Exmoor pony in its native habitat. (Photo Wolfgang Frey)

9 Outlook

With the realisation that the Exmoor pony is a surviving European mountain pony, the idea of an ancient breed of domestic horse that has been cultivated for 200 years collapses. The Exmoor pony is not a breed of horse, as the term breed is restricted to domestic animals. Treating it as such, with selective typological narrowing of variability, stallion licensing and similar subjective measures, leading to livestock genesis in the long run, should make way to a conservation strategy like for any other endangered species. The original English Exmoor Pony Society and its subsidiary organisations outside Great Britain are by no means superfluous. On the contrary, it will take on a more far-reaching additional significance as a species conservation organization if it is prepared to take on this task and thus changes its statutes.

This society, founded for the conservation of the Exmoor pony, became responsible for a counterproductive breeding control with the knowledge of the time in the interest of keeping a domestic horse breed typologically pure, but not of preserving the diversity of a wild animal population. In the long term, it would massively jeopardise the continued existence of the original Exmoor pony, which has survived as a wild horse into the 21st century, and prevent its survival as a wild horse in the long term if this path were to be pursued further.[125]

It makes sense that, similar to the Przewalski's horse, two conservation or breeding lines should be realised side by side. The current breeding policy of the Exmoor Pony Society and its various national subsidiary organisations is fundamentally incompatible with the requirements of conserving a wild animal. However, it is needed, not least in the commercial interest of the breeders, in the interest of the many Exmoor pony owners and friends and in the interest of safeguarding genetic resources. Therefore, there should not be an *either – or*, but a *both – and*, in which both requirements are taken into account.

Conservation breeding means a wild pony line in connection with the International Union for Conservation of Nature (IUCN), with

---

[125] Willman, R. (1999): Das Exmoor-Pferd: eines der ursprünglichsten halbwilden Pferde der Welt. Natur und Museum, 129, 12: 389-407.

as many free-range populations as possible. Stallions may be exchanged between them from time to time, if they are mini-populations with only one or two harem groups. In principle, however, they remain completely unaffected without any breeding control beyond occasional removal for population regulation. In the interest of natural selection pressure, the establishment of such populations in areas colonised by large carnivores, bears and wolves, as natural regulators, is desirable. A combination of conservation breeding and landscape management seems justifiable. European mountain regions appear particularly suitable for such conservation measures. Here, the original association with red deer and possibly ibex, possibly even European bison, could be realised.

In contrast to, but parallel to, conservation breeding is *purpose breeding*. This term is used here to describe the continuation by the Exmoor Pony Society of the breeding procedure that has been in use for a century. By means of stallion approvals and breeding exclusions of individuals that do not correspond to the subjectively defined image of the animals, it maintains the appearance in a uniform manner, more uniform than would be expected in a wild population, but steers the character towards less aggressiveness ("wildness") or towards docility, i.e. towards livestock.[126] The breeding process of the B-line of the Przewalski's horse was initially similar to this procedure.

The term purpose breeding was chosen because it emphasises the practical purposes of horse breeding, i.e. riding, driving and marketing. It goes without saying that the farmers on Exmoor, whose families have ensured the preservation of the Exmoor pony for two centuries by using the animals, must not be disadvantaged in the slightest by the realisation of its wild horse nature. If anything, this realisation should be used to develop a stronger nature tourism industry, which may bring economic revitalisation to the region.

---

[126] Willman, R. (1999): Das Exmoor-Pferd: eines der ursprünglichsten halbwilden Pferde der Welt. Natur und Museum, 129, 12: 389-407.
Baker, S. (2017): Exmoor Pony Chronicles. Wellington (Halsgrove).

Figure 9.1: Exmoor pony mare with foal (Equus ferus arcelini). The British wild horse, which has only survived as a result of a two-hundred-years-old misinterpretation (a very special breed of domestic horses worthy of protection, but widely misunderstood in terms of its true nature), should be given a free future with as little interference as possible as a highly significant British and pan-European natural heritage. (Photo Wolfgang Frey in Exmoor National Park)

10 Register

Adaptation   17, 19, 24
Addax   23
Addax nasomaculatus   23
Akhal-teke horse   52
A-line   47, 63 ff., 67 f.
Allen's rule   38
Allometry   28, 73, 75
Alpaca   22
Asinus   33
Ass   33
Attentiveness   67, 69
Aurignacian   72
Aurochs   72

Behaviour   17 ff., 25, 30, 57,
    61 ff.
Bible   18
Biological species concept   32,
    35
Bison   71 f., 79, 81, 90
Bison bison   81
Bison priscus   71
B-line   47 f., 63 ff., 67, 78, 90
Botai culture   42
Brain   19, 28 ff., 47, 70 ff., 84
Brand   5
Breed   12 ff., 50, 52, 54, 68, 87,
    89, 91
Broad-footed horse   40, 42, 45,
    50, 73 f., 78 ff., 81
Bronze Age   14, 26, 54,

Capra pyrenaica   81
Cave painting, art   7, 80, 82

Celtic-Germanic breed group
    50, 73, 77 f., 84
Celtic pony   50, 73 f., 78
Cephalisation   27, 30 ff.
Cervus elaphus   22 f., 81
Character   8, 78
Coat   5, 78, 80, 84
Cold-blooded horse   53, 86
Colour   4 ff., 24 ff., 56 ff., 79
    f., 84 ff.
Connemara horse   50
Coordination   61, 65 f., 69
Crossbreeding   10, 46 f., 57, 67

Danube   43
Dartmoor-Pony   11, 57, 66 ff.
Desna   39
Dnipro   43
Dolichohippus   33
Domestication   7, 17 ff., 22, 24,
    26 ff., 40, 42, 44, 87
Domestic horse   7, 13 ff., 22,
    29 f. 33 ff., 41 f., 47, 50
    ff., 57, 65, 67 f., 70 ff., 76
    ff., 83 f., 86, 89, 91
Don   39, 43 ff., 48, 50
Dülmener horse   50, 57
Dwarf pony   53
Dzungarian wild horse, see
    Przewalski's horse

Ear   5
Eel line   7, 56
Eisenmann horse type   39, 77
Elephant   24, 84

Elephas maximus 24
Equus 33 ff.
Equus africanus 34
Equus asinus 34
Equus caballus 33 ff., 77
Equus ferus 33 ff., 43
Equus ferus antunesi 40, 73 ff.,
    76 f., 79
Equus ferus arcelini 40, 73 f.,
    76 ff., 91
Equus ferus ferus 33, 40, 43
    ff., 79
Equus ferus gallicus 40, 74, 76
    f.
Equus ferus germanicus 40, 74
    f., 76 f., 80
Equus ferus gmelini 45, 73, 75,
    77, 79
Equus ferus latipes 40, 73
Equus ferus lenensis 42
Equus ferus mosbachensis 38
Equus ferus przewalskii 33, 42,
    45 ff., 79
Equus ferus silvestris 45, 79
Equus grevyi 34
Equus hemionus 23, 34
Equus quagga 34
Equus zebra 34
European mountain pony 78,
    80 ff., 86, 89
Exmoor-Pony Society 9, 89 f.
Eye rim, ring 4 f., 80

Fallow deer 26, 87
Family 32
Fjord horse 62
Forelock 5, 80
Forest-steppe 40, 43 f.

Forest tarpan 45, 57 f., 79, 83
Franco-Cantabrian mountains
    72, 82
Friesian horse 50

Gallo-roman horse 75
Game 10, 12, 18, 20 f., 35, 54
    f., 61, 65, 69, 72, 78 f., 83
Game farming 23
Garrano horse 56
Genus 32 f.
Gotland horse 58
Grevy's zebra 33 f.
Guanaco 21 f.

Habitat 8 f.
Haplotype 70
Heck horse 58 f., 63, 65, 67
Hemione 23, 33 f.
Hemionus 33 f.
Hippological horse types see
    horse types
Hippotigris 33 f.
Horse types 13, 15, 39 f., 44,
    56, 73
Hybridisation 55, 71

Ibex 72, 80, 90
Icelandic horse 11, 26, 50, 58,
    62, 66 ff., 84 f.
Improvement 10, 12, 57, 87
Inca 21 f.
Individual distance 62 ff.
Introgression 45, 47, 70
Iron Age 73, 78
Isolation mechanism 31 f., 35

Konik 57 ff., 63

La Pasiega 72
Late Pleistocene 38 ff., 43, 70
    f., 73
Livestock 9 f., 12, 17 ff., 25
    ff., 31, 34 f., 42, 48, 50 ff.,
    60 ff., 65, 68 ff., 78 f., 81,
    83 f., 87, 90
Llama 22

Magdalenian 39, 72, 77, 80
Mane 5, 72
Marismeño horse 55
Mealy muzzle 4 ff., 80
Metacarpus 73, 78
Metapodium 13, 38, 39 f., 80
Metapopulation 31 ff., 72
Middle Pleistocene 38, 42
Mitochondrial DNA 15, 70
Molar 39, 72, 76 f., 80
Molecular genetics 15 f., 19 f.,
    26, 36, 42 f., 47, 57, 70, 83
    f., 86 f.
Mongolian horse 47
Mouflon 19, 30 f.
Mountain zebra 33 f.
Movement intensity 68 f.
Mustang 54

Namib pseudo-wild horse 54
Neolithic 26,30, 41
Neumühle-Riswicker Hirsche
    87
New Forest-Pony 11, 57
Niaux 77, 80, 82
Nomenclature 31, 33, 35, 43 f.,
    79

North-west European breed
    group 50, 62, 65, 74, 86

Oriental breed group 50, 62
Oryx 23
Oryx gazella dammah 23

Phylogenetic species concept
    32, 35
Piktish horse 50
Pleistocene 26, 38, 40, 43, 73
    f. , 77 f.
Population 16, 19, 25, 31 f.,
    40, 42, 45, 54, 57, 70 ff.,
    80 f., 83 f., 86 f., 89 f.
Post-mating isolation
    mechanism 31, 54
Preadaptation 24 ff., 84
Pre-mating isolation mechanism
    31, 35, 39, 54
Premolar 39, 72, 76 f., 80
Principle of priority 33 f.
Protocone 39 f., 76 f.
Przewalski's horse 7, 14 f., 28
    ff., 33, 42 f., 45 ff., 50, 58,
    63, 65, 67 ff., 78, 89 f.
Pseudo-wild horse 54 ff.
Psychophysical correlation 25

Quagga 33 f.
Qur'an 18

Rabbit 19 f.
Red deer 22 f., 72, 81 f. 90
Relict 15, 41, 80, 83, 87
Russian tarpan 33, 44 f., 79

Sheep 19, 30 f., 82

Shetland pony 50
Shoulder height 4
Shyness 8, 46
Sorraia horse 50, 56, 62, 65
Species 31 ff., 39 f., 44 f.
Speed/Ebhardt horse types see
    Horse types
Standard-bred horse 62
Steppe 37, 40, 43 f., 73
Steppe horse 40, 73, 84, 86
Steppe-tundra 40
Steppe zebra 25, 33 f.
Subgenus 32
Subspecies 32, 40, 42, 44 f., 75
    f., 80 f.
Synchronisation 65, 67
Systematics 31 f., 35

Tarpan 40 f., 43 ff., 49, 58 f.,
    85
Taxonomy 31 ff., 35, 39, 76,
    80 f.
Thorough-bred horse 50
Toad eye 4
Trakehner horse 68 f.
Tundra 37
Type specimen 43

Ukrainian tarpan 45, 71, 73, 75
    f., 79

Viking horse 50
Vikuña 21 f.
Volga 43 ff., 50

Warm-blooded horse 50, 62,
    65

Wild horse 13 ff., 26, 30, 33,
    36 f., 38 ff., 49 f., 54, 65,
    70 ff., 79, 81, 83 f., 86, 89
    ff.
William the Conquerer 9